Handbook of
Geometrical Tolerancing

Handbook of
Geometrical Tolerancing
Design, Manufacturing and Inspection

G. Henzold
Ratingen, Germany

JOHN WILEY & SONS
Chichester · New York · Brisbane · Toronto · Singapore

Other Wiley Editorial Offices

John Wiley & Sons, Inc., 605 Third Avenue,
New York, NY 10158-0012, USA

Jacaranda Wiley Ltd, 33 Park Road, Milton,
Queensland 4064, Australia

John Wiley & Sons (Canada) Ltd, 22 Worcester Road,
Rexdale, Ontario M9W 1L1, Canada

John Wiley & Sons (SEA) Pte Ltd, 37 Jalan Pemimpin #05-04,
Block B, Union Industrial Building, Singapore 2057

Reprinted May 1997

British Library Cataloguing in Publication Data

A catalogue record for this book is available from the British Library

ISBN 0 471 94816 0

Typeset in 9/11pt Times by Lasertext Ltd., Stretford, Manchester
Printed in Great Britain by Antony Rowe Ltd, Chippenham, Wiltshire

Contents

Contents

Contents

Contents

Preface

Technical developments leading to higher performance, higher efficiency, lower pollution and greater reliability, and the constraints imposed by economy and the need for rationalization, greater cooperation with licensees and subcontractors, make it necessary to have completely toleranced drawings suitable for the function, for manufacturing and inspection.

"Completely toleranced" means that the geometries (form, size, orientation and location) of the geometrical elements of a workpiece are completely defined and toleranced. Nothing can be left to the individual judgement of the manufacturer or inspector. Only with this approach can proper functioning be verified and the possibilities available for economization and rationalization in manufacturing and inspection be fully utilized.

This book presents the state of the art regarding geometrical tolerancing. It describes the international standardization in this field, which is laid down in ISO Standards. It indicates the deviations between the American National Standard ANSI Y14.5M, the East European Standards and the ISO Standards. It describes the additional specifications laid down in the British Standard BS 308 and the German Standards (DIN Standards). Possible further developments in the field of geometrical tolerancing are also pointed out.

Principles for the inspection of geometrical deviations are given, together with a basis for tolerancing suitable for inspection.

Examples of tolerancing appropriate to various functional requirements, a guide for geometrical tolerancing, are described.

This book may serve as an introduction to geometrical tolerancing for students, and may also help practitioners in the fields of design, manufacturing and inspection.

The author has used his experiences gathered during the elaboration of the ISO Standards and during his lectures and discussions on geometrical tolerancing in industry and education. However, proposals for improvements are appreciated, and should be directed to the publisher.

The book represents the author's understanding of the various standards, and is written for educational purposes. In cases of dispute the original standards should be consulted.

Georg Henzold is manager of the department for standardization of a manufacturer of power plant machinery. He is chairman of the committees dealing with standardization in the field of geometrical tolerances in the German Standardization Institut (DIN) and in the European Committee for Standardization (CEN), and is a long-time delegate in the pertinent committees of the

International Standardization Organization (ISO).

In this book he describes the aspects of geometrical tolerancing on the basis of the ISO Standards. He gives the necessary background information, various examples of applications in industry, and a comparison with the American National Standard ANSI Y14.5 M, the British Standard BS 308, the German Standards (DIN Standards) and East European Standards.

Notation

A reading, measured value
B zone
C mean size
H height
I actual size
L length
Ma maximum material size (MMS)
Mi least (minimal) material size (LMS)
N number
P mating size or content or statistically probability or pitch
R calculated value from readings
T size tolerance
Va maximum material virtual size (MMVS)
Vi least (minimum) material virtual size (LMVS)
Z centre

a deviation or variation
b width
c distance or safety factor with statistical tolerancing
d diameter
e local deviation
h height
k correction factor
l length
n number
p coordinate
q coordinate
r radius
s clearance
t geometrical tolerance
u measuring uncertainty
x coordinate
y coordinate
z coordinate

Δ difference
α angle

Notation

β angle
γ angle
δ deviation corresponding to ISO 1101
λ cut-off, wavelength

Subscripts

a   coaxiality or arithmetical
b   profile of a line
c   position or section
d   orientation
e   flatness or external
f   form
g   straightness
h   profile of a surface
i   internal
k   crossing
l run-out or left
m   measured value or centre
n   perpendicularity
o   location
ö   local
p   parallelism or point
q   longitudinal section profile
r   roundness or right
s   symmetry or statistical
t   total run-out
th   theoretically exact
u   orientation or location or below
v   twist
w   angularity
x x direction
y y direction
z z direction
z   cylindricity

Abbreviations

AVG	average
BASIC	theoretical exact dimension
DIA	diameter
D&T	dimensioning and tolerancing
FIM	full indicator movement
FIR	full indicator reading
FRTZF	feature-relating tolerance zone framework
GD&T	geometrical dimensioning and tolerancing
GPS	geometrical product specification
LD	least (minimum) diameter
LMC	least material condition
LMR	least material requirement
LMS	least material size
LMVC	least material virtual condition
LMVS	least material virtual size
LSC	least-squares circle
MCC	minimum circumscribed circle
MD	major diameter
MIC	maximum inscribed circle
MMC	maximum material condition
MMR	maximum material requirement
MMS	maximum material size
MMVC	maximum material virtual condition
MMVS	maximum material virtual size
MZC	minimum-zone circle
PD	pitch diameter
PLTZF	pattern-locating tolerance zone framework
RFS	regardless of feature size
SEP REQT	separate requirement
SIM REQT	simultaneous requirement
TIR	total indicator reading
TP	true position, theoretical exact position or location
VD&T	vectorial dimensioning and tolerancing

1 Properties of the Surface

The suitability of a workpiece for a given purpose depends on its internal properties (material properties, internal discontinuities such as shrink holes, and internal imperfections such as segregations) and its surface condition.

The surface condition comprises the properties of the surface border zone. These are chemical, mechanical and geometrical properties (Figure 1.1).

The chemical and mechanical properties comprise chemical composition, grain, hardness, strength and inhomogeneities. The properties of the surface border zone may be different from those of the core zone.

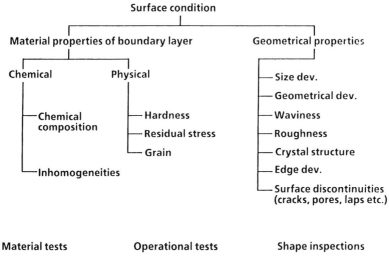

Figure 1.1: Properties of surfaces and their tests and inspections.

The geometrical properties are defined as deviations from geometrical ideal elements (features) of the workpiece. Geometrical ideal elements (features) are parts of the entire workpiece surface that have unique nominal geometrical forms (e.g. planes, cylinders, spheres, cones and tori). They can also be derived for example as axes, section lines, generator lines, lines of highest points and edges (Figure 1.2).

1

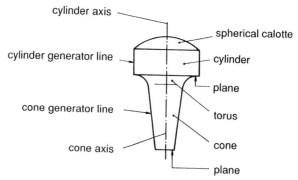

Figure 1.2: Examples of geometrical elements.

Geometrical deviations comprise

- size deviations
- form deviations
- orientational deviations
- locational deviations

- waviness
- roughness
- surface discontinuities
- edge deviations

Size deviation This is the difference between actual size and nominal size. A distinction has to be made between

- deviation of the actual local size from the nominal linear size or from the nominal angular size
- deviation of the actual substitute size from the nominal substitute size (see 8).

The actual local linear sizes are assessed by two-point-measurements (ISO 8015, ISO 286). The actual local angular sizes are assessed by angular measurements of averaged lines (ISO 8015, ISO 1947); see 3.7 (Figure 3.42).

The actual local sizes of the same geometrical element of one workpiece differ depending on their location. The actual substitute size is unique and therefore representative of the entire geometrical element of one workpiece.

Size deviations are assessed over the entire geometrical element. They originate mainly from imprecise adjustment of the machine tool and variations during the manufacturing process (e.g. due to tool wear).

Form deviation This is the deviation of a feature (geometrical element, surface or line) from its nominal form (Figure 1.3). If not otherwise specified, form deviations are assessed over (or along) the entire feature. Form deviations originate for example from the looseness or error in guidances and bearings of the machine tool, deflections of the machine tool or the workpiece, error in the fixture of the workpiece, hardness deflection or wear. The ratio between width and depth of local form deviations is in general more than 1000:1 (VDI/VDE 2601).

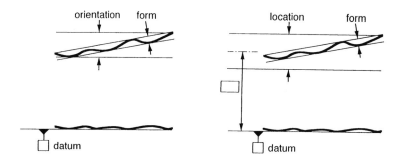

Figure 1.3: Form deviation, orientational deviation and locational deviation.

Orientational deviation This is the deviation of a feature from its nominal form and orientation. The orientation is related to one or more datum feature(s). The orientational deviation includes the form deviation (Figure 1.3). If not otherwise specified, orientational deviations are assessed over the entire feature. Orientational deviations originate similarly to form deviations. They also arise owing to erroneous fixture of the workpiece after re-chucking.

Locational deviation This is the deviation of a feature (surface, line or point) from its nominal location. The location is related to one or more datum feature(s). The locational deviation also includes the form deviation and the orientational deviation (of the surface, axis or symmetry face) (Figure 1.3). If not otherwise specified, locational deviations are assessed over the entire feature. Locational deviations originate similarly to size, form and orientational deviations.

Waviness This comprises more or less periodic irregularities of a workpiece surface with spacing greater than the roughness spacing (DIN 4774). The ratio between spacing and depth of the waviness is in general between 1000:1 and 100:1. In general more than one wave can be recognized (VDI/VDE 2601). The waviness is assessed from one or more representative parts of the surface (Figure 1.4). Waviness originates from eccentric fixture during the manufacturing process,

3

from form deviations of the cutter, or from vibrations of the machine tool and/or the cutting tool and/or the workpiece (DIN 4760). Waviness will not be dealt with in the following.

Roughness This comprises periodic or aperiodic irregularities of a workpiece surface with small spacings inherent to the forming process. The ratio between spacing and depth of the roughness is in general between 150:1 and 5:1 (VDI/VDE 2601). The roughness is assessed from one or more representative parts of the surface (Figure 1.4).

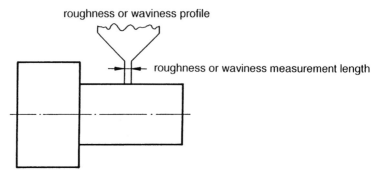

Figure 1.4: Assessment of roughness or waviness.

Roughness originates from the direct effect of the cutting edges (VDI/VDE 2601) i.e. from printing the cutting edges on the surface. As a result of the cutting process (tear chip or shear chip), the print will be modified. Other origins are deformations from blasting, gemmation with galvanizing, crystallization and chemical effects (mordants, corrosion). Roughness ranges down to the crystal structure (DIN 4760). Roughness will not be dealt with in the following.

Surface discontinuity This is an isolated cut of the surface, like a crack, pore or lap. In general it is not taken into account when assessing deviations of size, form, orientation, location, waviness and roughness. The definitions and sizes and permissibilities of surface discontinuities have to be dealt with separately.

Up to now there have been very few standards on this subject available (e.g. for fasteners ISO 6157 and for hot-milled steel products ISO 9443). Surface discontinuities will not be dealt with in the following.

Edge deviations These are deviations of the workpiece edge zone from the geometrical ideal shape, like a burr or abraded edges instead of sharp edges. There is an International Standard ISO 13715 in preparation on terms and definitions and drawing indications for permissible edge deviations.

The **classification of surface irregularities** as described above is useful for the following reasons.

(a) The different kinds of surface irregularities have different origins in the manufacturing process. In order to control the manufacturing process, these must be assessed separately.

(b) The different kinds of surface irregularities often have different effects on the suitability of the surface for its purpose. For example, on raceways of ball bearings waviness has a great influence on lifetime and noise, while roughness has only little influence. In order to specify the permissible function-related deviations, the different kinds must be specified separately.

(c) The depths of the irregularities vary over large ranges, between about 0.1 μm (and sometimes smaller) with roughness and about 100 μm (and sometimes more) with form deviations. The ratio between spacing and depth of the irregularities also varies greatly. The smallest ratio occurs with cracks, and is in general less than 5:1, whereas the ratio with form deviations is in general greater than 1000:1. Because of these wide ranges, the requirements for measuring devices and for diagrams are quite different. For the assessment of different kinds of irregularities (deviations), different kinds of measuring instruments with different magnifications and different profile diagrams with different ratios of horizontal to vertical magnifications are used.

The definitions of the different kinds of irregularities (deviations) are rather uncertain. There are no distinct borderlines. Therefore it was discussed in ISO whether to define borderlines in terms of defined spacings of irregularities, in terms of defined ratios between spacings and depths of irregularities, in terms of defined ratios between spacings of irregularities and feature lengths. However, it was decided to retain the definitions according to the causes of the irregularities (ISO 4287, ANSI B46.1, BS 1134 , DIN 4760).

There is another distinction, namely that between **micro- and macro-deviations**. Macro-deviations are those that can be assessed with the usual measuring devices for the assessment of size, form, orientation and location (e.g. a dial indicator). Micro-deviations are those that are assessed with roughness- or waviness-measuring instruments. Macro-deviations are assessed over the entire feature length; micro-deviations are assessed from a representative part of the surface. Here also there is no distinct borderline, because sometimes parts of the waviness will contribute to the result of the measured macro-deviations, while sometimes parts of the form deviations will contribute to the result of the measured micro-deviations (waviness).

Properties of the Surface

Figure 1.5 gives an idea of the combination (superposition) of the kinds of surface irregularities (deviations) on a surface.

Geometrical deviation Profile diagram	Description Examples of origin
1st order: Form 	errors in guidance of machine tool, deflections of machine tool or workpiece, error in fixture of workpiece, warping, wear
2nd order: Waviness 	eccentric fixture, form deviation of tool, vibration
3rd order: Roughness 	grooves; form of tool cutting edge, horizontal and vertical feed
4th order: Roughness 	cutting process (tear chip, shear chip), deformation from blasting, gemmation with galvanizing
5th order: Roughness not presentable	crystallization process, mordant, corrosion
6th order: Roughness not presentable	crystal structure
Superposition 	actual surface

Figure 1.5: Superposition of surface deviations (DIN 4760).

2 Principles of Tolerancing

It is impossible to manufacture workpieces without deviations from the nominal shape. Workpieces always have deviations of size, form, orientation and location.

When these deviations are too large, the usability of the workpiece for its purpose will be impaired. When during manufacturing attempts are made to keep these deviations as small as possible, in order to avoid impairing usability, in general the production is too expensive and the product is hard to sell.

In general, competition forces the use of all possibilities for economic production, including those arising from recent developments. Therefore it is necessary that the drawing tolerances the part completely, i.e. each property (size, form, orientation and location) must be toleranced. Only then is the manufacturer able to choose the most economic production method, e.g. depending on the number of pieces to be produced and on the production methods available.

Incompletely toleranced drawings result in

- questions for the production planning engineer,
- questions for the manufacturing engineer,
- questions for the inspection engineer,
- reworking,
- defects,
- damages.

Only completely toleranced drawings enable the production of workpieces that are as precise as necessary and as economic as possible. This is necessary to ensure competitive ability.

When all tolerances necessary to define the workpiece completely are indicated individually, the drawing becomes overloaded with indications and is hard to read. Therefore general tolerances have been standardized (e.g. ISO 2768) that should be applied when no individual (in general, smaller) tolerance is indicated.

General tolerances should be equal to or larger than the customary workshop accuracy. The customary workshop accuracy is equal to the tolerances that the workshop does not exceed with normal effort using normal workshop machinery. Larger tolerances bring no gain in manufacturing economy. The normal workshop accuracy depends on the workshop machinery that produces the largest deviations (disregarding exceptions, which are to be dealt with exceptionally). The customary workshop accuracy is in general the same within one field of industry. For example, the customary workshop accuracy for material removal in the machine-building industry corresponds to the general tolerances ISO 2768 - mH.

The general tolerances should be applied by indicating in the title box of the drawing a reference to the applicable standard and tolerance class.

Tolerances that must be smaller have to be indicated individually (see 16). This concept is only applicable where appropriate standards on general tolerances are available. Where such standards do not exist, the following has to be done:

- refer to a company standard on appropriate general tolerances,

- indicate all tolerances necessary to define the geometry (size, form, orientation, location), or

- indicate all tolerances that are considered as possibly being exceeded by the planned manufacturing process.

The last method provides drawings that are not completely toleranced and that are therefore incomplete and may lead to the disadvantages described above.

3 Principles of Geometrical Tolerancing

3.1 SYMBOLS

The symbols for the drawing indications of geometrical tolerances according to ISO 1101, ISO 5459, ISO 286 and ISO 10579 are shown in Tables 3.1 and 3.2.

Table 3.1: Symbols for geometrical tolerancing.

Symbol	Interpretation
// ∅ 0,02 A	Additional indications (e.g. number of features)
	Tolerance frame
	Additional indications (e.g. quality of feature)
	Datum letter with related geometrical tolerances
	Tolerance value
	Symbol for cylindrical tolerance zone
	Symbol for geometrical tolerance
4x	Number (4) of toleranced features
	Arrow line for indication of the toleranced feature
A	Arrow with letter for indication of the toleranced feature
A	Datum triangle and datum frame for indication of the datum
●	Continuous leader line, pointing to the toleranced element or datum that is the visible surface
● - - - -	Dashed leader line, pointing to the toleranced element or datum that is the hidden surface

Table 3.1 (cont.)

Symbol	Interpretation
	Axis or median face as toleranced and datum feature
	Generator line or surface as toleranced and datum feature
20	Theoretical exact dimension
	Restricted part of the feature (toleranced or datum)
− | 0,1/100	Restricted length of the toleranced feature lying anywhere
Ⓔ	Envelope requirement
Ⓜ	Maximum material requirement
Ⓛ	Least material requirement
Ⓡ	Reciprocity requirement
Ⓟ	Projected tolerance zone
Ⓕ	Flexible part in free state

Table 3.1 (cont.)

Symbol	Interpretation
	Datum target
	Datum target size
	Datum target feature and datum target number
	Datum target point
	Datum target area ($\varnothing$ 4)
	Datum target line, front view
	Datum target line, side view

Table 3.2 shows the symbols for the geometrical tolerance characteristics. With the symbols in the left-hand column of Table 3.2, all kinds of geometrical tolerances can be expressed. The symbols in the right-hand column cover special cases that occure very frequently. These symbols are standardized in addition, and should be used in order to facilitate readability of drawings.

The proportions of the symbols for the drawing indications are standardized in ISO 7083. However, some CAD systems deviate from this standard. This may be tolerable as long as the drawing indications are unambiguous (as demonstrated in this book).

Principles of Geometrical Tolerancing

Table 3.2: Geometrical tolerances and tolerance symbols

Unrelated geometrical tolerances (form tolerances)

Profile (form) of lines†	⌒
Straightness	—
Roundness	○
Profile (form) of surfaces†	⌒
Flatness	▱
Cylindricity	⌔

Related geometrical tolerances

Orientation

Angularity	∠
Parallelism	//
Perpendicularity	⊥

Location

Position	⊕
Coaxiality	◎
Symmetry	≡

Run-out

Circular run-out	↗
Circular radial run-out	
Circular axial run-out	
Total run-out	⌗
Total radial run-out	
Total axial run-out	

†The tolerances of profiles of lines and of surfaces may also be related to datums (Figure 4.2).

3.2   DEFINITIONS OF GEOMETRICAL TOLERANCES

In order to specify geometrical tolerances, the workpiece is considered to be composed of features (geometrical elements), such as planes, cylinders, cones, spheres and tori (Figure 1.2).

Form tolerance   This is the permitted maximum value of the form deviation (see 18.6 and 18.7.1–18.7.5). According to ISO 1101, there are defined form tolerance zones within which all points of the feature must be contained. Within a zone, the feature may have any form, if not otherwise specified. The tolerance value defines the width of this zone (Figure 1.3).

Form tolerances limit the deviations of a feature from its geometrical ideal line or surface form. Special cases of line forms with special symbols are straightness and roundness (circularity) (Figure 3.1). Special cases of surface forms with special symbols are flatness (planarity) and cylindricity (Figure 3.2).

Orientation tolerance   This is the permitted maximum value of the orientation deviation (see 18.6 and 18.7.8). According to ISO 1101, there are defined orientation tolerance zones within which all points of the feature must be contained. The orientation tolerance zone is in the geometrical ideal orientation with respect to the datum(s). The tolerance value defines the width of this zone (Figure 1.3).

Orientation tolerances limit the deviations of a feature from its geometrical ideal orientation with respect to the datum(s). Special cases of orientation with special symbols are parallelism (0°) and perpendicularity (90°) (Figure 3.3).

The orientation tolerance also limits the form deviation of the toleranced feature, but not of the datum features(s). If necessary, a form tolerance of the datum feature(s) must be specified.

Location tolerance   This is twice the permitted maximum value of the location deviation (see 18.6 and 18.7.9). According to ISO 1101, there are defined location tolerance zones within which all points of the feature must be contained. The location tolerance zone is in the geometrical ideal orientation and location with respect to the datum(s). The tolerance value defines the width of this zone (Figure 1.3).

Location tolerances limit the deviations of a feature from its geometrical ideal location (orientation and distance) with respect to the datum(s). Special cases of location with special symbols are coaxiality (when toleranced feature and datum feature are cylindrical) and symmetry (when at least one of the features concerned is prismatic) where the nominal distance between the axis or median plane of the toleranced feature and the axis or median plane of the datum feature is zero (Figure 3.4).

The location tolerance also limits the orientation deviation and the form deviation of the toleranced feature (plane surface, axis or median face), but not

13

the form deviation of the datum feature(s). If necessary, a form tolerance for the datum feature(s) must be specified.

Run-out tolerances   These are partly orientation tolerances (axial circular run-out tolerance, axial total run-out tolerance) and partly location tolerances (radial circular run-out tolerance, radial total run-out tolerance). However, according to ISO 1101, they are considered as separate tolerances with separate symbols, because of their special measuring method.

Typical drawing indications and the relevant geometrical tolerance zones of tolerances of form, orientation, location and run-out are shown in Figures 3.1–3.9.

Which geometrical tolerances are applicable for which type of features are shown in Table 3.3. The table also shows the possible combinations of toleranced feature and datum feature.

The possible tolerance zones, their shapes, their orientations and locations, their widths and their lengths are described in 3.3.

The possibilities of prescribing datums are described in 3.4.

The definitions of axes and median faces are dealt with in 3.5.

The differences between angularity tolerances according to ISO 1101 and angular dimension tolerances according to ISO 8015 are described in 3.6.

Form tolerances of lines   These are shown in Figure 3.1.

Profile tolerances of a line (Figure 3.1 top)   The nominal (theoretical, geometrical ideal) line is defined by the theoretical exact dimensions. In each section, parallel to the plane of projection in which the indication is shown, the profile line should be contained between two equidistant lines enveloping circles of diameter 0.02, whose centres are located on a line having the nominal (theoretical, geometrical ideal) form (see also 4).

Roundnesss (circularity) tolerance (Figure 3.1 centre)   In each cross-section of the conical surface the profile (circumference) should be contained between two coplanar concentric circles with a radial distance of 0.02.

Straightness tolerance (Figure 3.1 bottom)   In each section, parallel to the plane of projection in which the indication is shown, the profile should be contained between two parallel straight lines 0.03 apart.

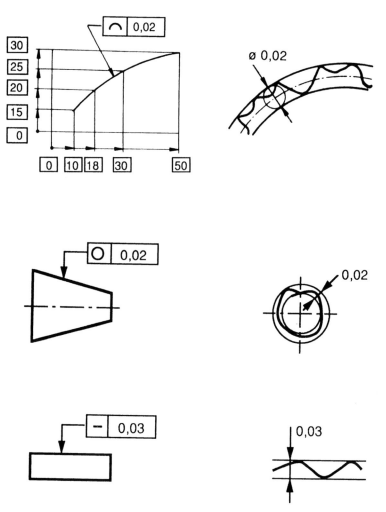

Figure 3.1: Form tolerances of lines, drawing indications and tolerance zones.

For further examples of form tolerances of lines see 4, 9, 11, 12, 20.1.2, 20.2, 20.3, 20.8 and 20.9.

Principles of Geometrical Tolerancing

Form tolerances of surfaces   These are shown in Figure 3.2.

Profile tolerance of a surface (Figure 3.2 top)   The nominal (theoretical, geometrical ideal) surface is defined by the theoretical exact dimensions. The surface should be contained between two equidistant surfaces enveloping spheres of diameter 0.03, whose centres are located on a surface having the nominal (theoretical, geometrical ideal) form (see also 4).

Cylindricity tolerance (Figure 3.2 centre)   The surface should be contained between two coaxial cylinders with a radial distance 0.05.

Flatness tolerance (Figure 3.2 bottom)   The surface should be contained between two parallel planes 0.05 apart (see also 20.2).
   For tolerancing of cones see 5.
   For further examples of form tolerances see 20.1.2, 20.3, 20.8 and 20.9.

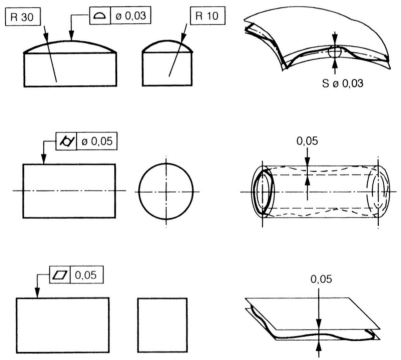

Figure 3.2:   Form tolerances of surfaces, drawing indications and tolerance zones.

3.2 Definitions of geometrical tolerances

Orientation tolerances   These are shown in Figure 3.3.

Angularity tolerance (Figure 3.3 top)   The actual axis should be contained between two parallel planes 0.1 apart that are inclined at the theoretically exact angle 60° to the datum A.

Perpendicularity tolerance (Figure 3.3 centre)   The surface should be contained between two parallel planes 0.08 apart that are perpendicular to the datum A.

Parallelism tolerance (Figure 3.3 bottom)   The surface should be contained between two parallel planes 0.08 apart that are parallel to the datum A.
   For further examples of orientation tolerances see 3.3, 20.1.1 and 20.8

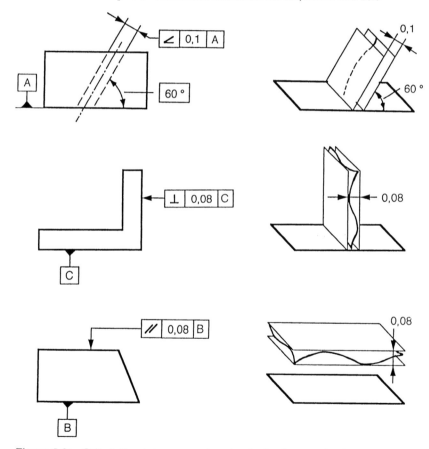

Figure 3.3:   Orientation tolerances, drawing indications and tolerance zones.

Location tolerances These are shown in Figure 3.4.

Positional tolerances (Figure 3.4 top) The theoretically exact (nominal) position is defined by the theoretical exact dimensions with respect to the datums A, B and C. The actual axis should be contained within a cylinder of diameter 0.05, whose axis coincides with the theoretically exact position.

Coaxiality tolerance (Figure 3.4 centre) The actual axis should be contained within a cylinder of diameter 0.03 coaxial with the datum axis A. When the features are practically two-dimensional (thin sheet, engraving), the tolerance is also referred to as the concentricity tolerance.

Symmetry tolerance (Figure 3.4 bottom) The actual median face should be contained between two parallel planes 0.08 apart that are symmetrically disposed about the datum median plane A.

For further examples of location tolerances see 3.3, 6, 7, 9, 11, 19 and 20.

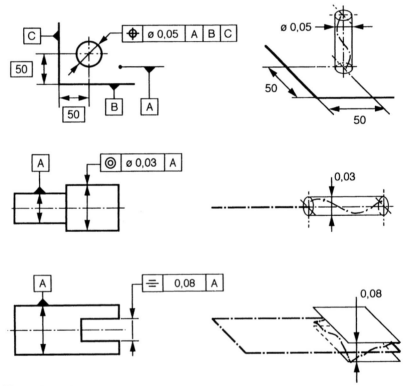

Figure 3.4: Location tolerances, drawing indications and tolerance zones.

Radial run-out tolerances   These are shown in Figure 3.5.

Circular radial run-out tolerance (Figure 3.5 top)   In each plane perpendicular to the common datum axis A–B the profile (circumference) should be contained between two circles concentric with the datum axis A–B and with a radial distance of 0.1.

Total radial run-out tolerance (Figure 3.5 bottom)   The surface should be contained between two cylinders coaxial with the datum axis A–B and with a radial distance of 0.1.

During checking of the circular radial run-out deviation, the positions of the dial indicator are independent of each other. However, during checking of the total radial run-out deviation, the positions are along a guiding (straight) line parallel to the datum axis A–B (see 18.7.10.3).

Therefore the straightness deviations and the parallelism deviations of the generator lines of the toleranced cylindrical surface are limited by the total radial run-out tolerance, but not by the circular radial run-out tolerance (Figure 3.6).

See also Figure 20.59.

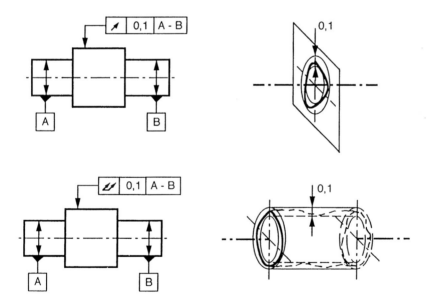

Figure 3.5:   Circular radial run-out tolerance, total radial run-out tolerance, drawing indications and tolerance zones.

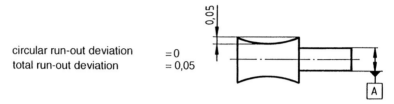

circular run-out deviation   = 0
total run-out deviation   = 0,05

Figure 3.6:   Workpiece with total radial run-out deviation but without circular radial run-out deviation.

Axial run-out tolerances   These are shown in Figure 3.7.

Circular axial run-out tolerance (Figure 3.7 top)   In each cylinder (cylindrical section, measuring cylinder) coaxial with the datum axis D the section line should be contained between two circles 0.1 apart and perpendicular to the datum axis D.

Total axial run-out tolerance (Figure 3.7 bottom)   The surface should be contained between two parallel planes 0.1 apart and perpendicular to the datum axis D.

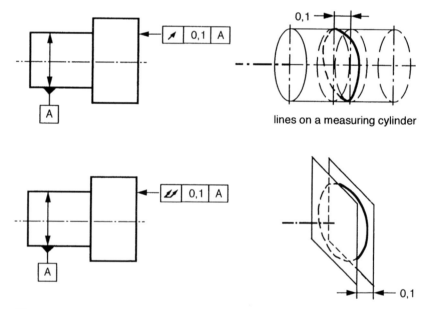

lines on a measuring cylinder

Figure 3.7:   Circular axial run-out tolerance, total axial run-out tolerance, drawing indications and tolerance zones.

During checking of the circular axial run-out deviation, the positions of the dial indicator are independent of each other. However, during checking of total axial run-out deviation, the positions are along a guiding (straight) line perpendicular to the datum axis D (see 18.7.8.3).

Therefore the flatness diviations of the tolerance surface are limited by the total axial run-out tolerance, but not by the circular axial run-out tolerance (Figure 3.8).

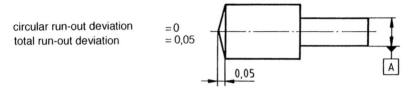

circular run-out deviation   = 0
total run-out deviation   = 0,05

Figure 3.8: Workpiece with total axial run-out deviation but without circular axial run-out deviation.

Run-out tolerances in any direction   These are shown in Figure 3.9.

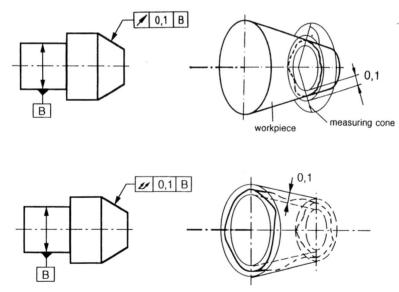

Figure 3.9: Circular run-out tolerance in any direction, total run-out tolerance in any direction, drawing indications and tolerance zones.

21

Principles of Geometrical Tolerancing

Circular run-out tolerance in any direction (Figure 3.9 top)   In each conical section (measuring cone) coaxial with the datum axis B and perpendicular to the nominal toleranced surface (defining the measuring cone angle) the section line should be contained between two circles 0.1 apart and coaxial to the datum axis B.

Total run-out tolerance in any direction (Figure 3.9 bottom)   The surface should be contained between two cones coaxial with the datum axis B and with a radial distance of 0.1 (measured perpendicular to the nominal cone surfaces). See also Figure 5.9.

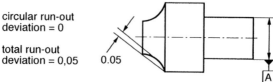

circular run-out
deviation = 0

total run-out
deviation = 0,05   0.05

Figure 3.10: Workpiece with total run-out deviation but without circular run-out deviation.

During checking the circular run-out deviation in any direction, the positions of the dial indicator are independent of each other. However, during checking of the total run-out deviation in any direction, the positions are along a guiding line (theoretically exact generator line of the toleranced future) parallel to its theoretically exact position with respect to the datum axis B.

Therefore the deviations of the generator line of the toleranced feature are limited by the total run-out tolerance in any direction, but not by the circular run-out tolerance in any direction.

Coaxiality tolerance and radial run-out tolerance are different. The former assesses the deviation of the axis from the datum axis, while the latter assesses the deviation of the circumference line from a coaxial circle. The radial run-out deviation is composed of the coaxiality deviation and parts of the roundness deviation. For limiting imbalance, for example, the coaxiality tolerance is appropriate to the function rather than the radial run-out tolerance. Figure 3.11

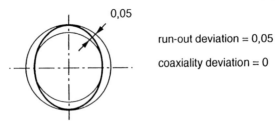

0,05

run-out deviation = 0,05

coaxiality deviation = 0

Figure 3.11: Workpiece with radial run-out deviation, but practically without coaxiality deviation.

22

shows an example where there is a radial run-out deviation but practically no coaxiality deviation.

Possibilities of geometrical tolerancing of features are listed in Table 3.3. It indicates which features may be form-toleranced and which may serve as toleranced feature and datum feature for geometrical tolerancing.

Table 3.3: Possible tolerancing of features (form elements). *The Drawing indication "section lines" or "lines only" is necessary (Figure 20.8). This indication may be omitted when the drawing already shows that only section lines can be applied (Figures 20.7 and 20.9).

	Tolerance	Symbol	Toleranced feature					Datum feature				
			Section line	Edge	Axis	Median face	Surface	Section line	Edge	Axis	Median face	Surface
Unrelated tolerance	Line profile	⌒	X	X	X							
	Straightness	—	X	X	X							
	Roundness	○	X	X								
	Surface profile	⌒				X	X					
	Flatness	▱				X	X					
	Cylindricity	⌀					X					
Related tolerance	Line profile with datum(s)	⌒	X	X				X	X	X	X	X
	Surface profile with datum(s)	⌒				X	X	X	X	X	X	X
	Inclination	∠	X*	X	X	X	X	X*	X	X	X	X
	Parallelism	//	X*	X	X	X	X	X*	X	X	X	X
	Perpendicularity	⊥	X*	X	X	X	X	X*	X	X	X	X
	Position	⊕		X	X	X	X		X	X	X	X
	Coaxiality	◎			X					X		
	Symmetry	≡			X	X				X	X	
	Circular run-out	⟋	X	X						X		
	Total run-out	⟋⟋		X			X			X		

For **Limitation of deviations by geometrical tolerances of higher order** see Figure 1.3.

Related geometrical tolerances (tolerances of orientation and location according to ISO 1101 and ANSI Y14.5M) define zones within which all points of the toleranced feature have to be contained. Therefore related geometrical tolerances limit also certain form deviations.

For example, related geometrical tolerances of surfaces also limit the form deviations of these surfaces (Figures 3.3 centre and bottom, 3.5 bottom, and 3.7 bottom).

Related geometrical tolerances of section lines limit the form deviations of the section lines but not the form deviations of the surface (in the other direction) (Figures 3.5 top and 3.7 top).

Related geometrical tolerances of axes or median faces (median planes) limit the form deviations of the axes or median faces, but not of the pertinent surfaces (Figures 3.3 tops and 3.4 bottom). (According to ISO 1101 and ANSI Y14.5 M, it is assumed that actual axes and actual median faces of workpieces are subject to form deviations (deviate from the geometrical ideal shape); see also 3.5.)

The indication of form tolerances is not necessary when the related geometrical tolerance already limits the form deviations to an sufficient extent.

According to East European Standards (former Comecon Standards), form deviations are not limited by related geometrical tolerances (see 21.2).

Similarly, locational tolerances (e.g. positional tolerances) limit the location but also the orientation and the form of the toleranced feature (e.g. axis and plane surface) (Figure 3.12); see also 8 and 18.10.

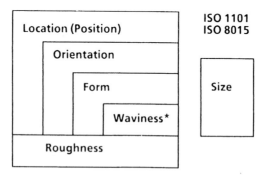

Figure 3.12: Assessment of types of geometrical deviations by geometrical tolerances (*ISO Standard under preparation).

3.3 TOLERANCE ZONE

3.3.1 Form of the tolerance zone

Depending on the toleranced characteristic and depending on the drawing indication the tolerance zone is one of the following:

- area within a circle (Figure 3.13);
- area between two concentric circles (Figure 3.14);
- area between two equidistant lines or between two parallel straight lines (Figure 3.15);
- space within a sphere (Figure 3.16);
- space within a cylinder (Figures 3.17 and 3.20);
- space between two coaxial cylinders (Figure 3.18);
- space between two equidistant faces or between two parallel planes (Figure 3.19);
- space within a parallelepiped (Figures 3.20 and 3.21) (see also text for Figure 6.2).

When the $\varnothing$ symbol precedes the tolerance value, the tolerance zone is cylindrical (Figures 3.17 and 3.20) or circular (Figure 3.13). In the absence of the $\varnothing$ symbol, the tolerance zone is limited by two lines (straight lines, circles or enveloping lines; Figures 3.14 and 3.15) or by two faces (planes, cylinders or enveloping faces; Figures 3.18–3.21), the distance between which is equal to the tolerance value and, according to ISO 1101, in the direction of the leader line arrow (see 3.3.3). For the tolerance the dimensions (mm or in) in the drawing apply.

For section lines the tolerance zones are two-dimensional (Figures 3.14 and 3.15). For thin parts the tolerance zone is practically two-dimensional (Figure 3.13). For edges, axes and median faces and surfaces the tolerance zone is three-dimensional (Figures 3.17–3.21).

For the straightness tolerance of generator lines of cylindrical or conical features the standards do not specify whether section lines in planes containing the axis * or lines of the highest points are applied. Mostly the function refers to the lines of the highest points. Therefore they are applied here (Figure 3.19).

For axes the tolerance zone is three-dimensional. When the tolerance is indicated in only one direction, the deviation is limited in only this direction (Figure 3.20

---

*Regression line or axis of the contacting cylinder or cone directed according to the minimum rock requirement or axis of the pair of coaxial cylinders or cones with least distance (see 3.5).

25

top).† When the tolerance value is preceded by the ⌀ symbol, the tolerance zone is cylindrical (Figure 3.20 bottom). When the tolerances are indicated in two directions perpendicular to each other for axes, the tolerance zone is a parallele-piped (Figures 3.20, 3.21 centre and 3.24). When the tolerance for an axis is indicated in two directions defined by two datums independent of each other (Figure 3.23), the tolerance zone does not necessarily have right-angles.

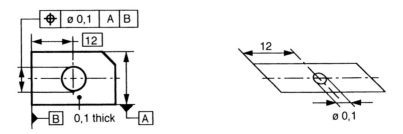

Figure 3.13: Circular tolerance zone.

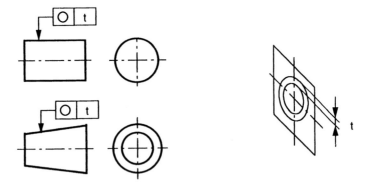

Figure 3.14: Tolerance zone between two concentric circles.

---

†In ISO 1101: 1983 the tolerance zone is sometimes shown as a zone projected onto a plane. According to Clause 13.3 of the standard, there is no difference in the meaning of tolerance zones shown projected onto a plane or shown three-dimensionally (Figure 3.22). In the next edition of ISO 1101 is intended to show the spatial (three-dimensional) tolerance zone only.

Figure 3.15: Tolerance zone between two parallel straight lines in a plane.

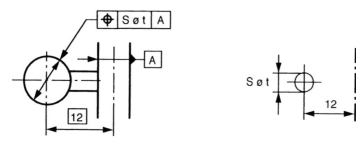

Figure 3.16: Spherical tolerance zone.

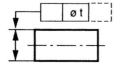

Figure 3.17: Cylindrical tolerance zone.

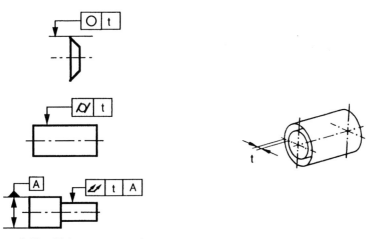

Figure 3.18: Tolerance zone between two coaxial cylinders.

27

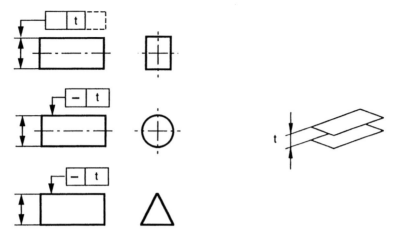

Figure 3.19: Tolerance zone between two parallel planes.

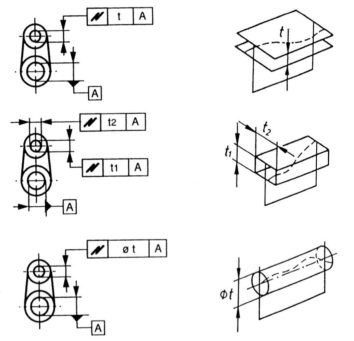

Figure 3.20: Spatial tolerance zone of a line.

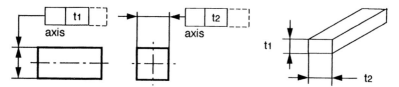

Figure 3.21: Parallelepipedal tolerance zone.

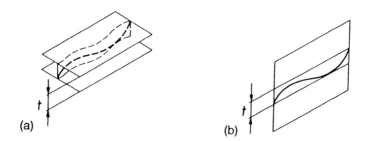

Figure 3.22: Tolerance zone of axes, (a) spatial representation, (b) projection of the spatial zone onto a plane (without any change in meaning).

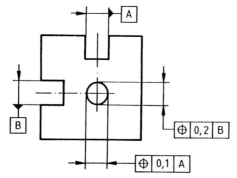

Figure 3.23: Tolerance zone with not necessarily right-angles (in section).

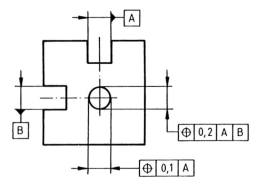

Figure 3.24: Tolerance zone with right-angles (in section).

3.3.2 Location and orientation of the tolerance zone

With **locational tolerances** and **run-out tolerances**, the tolerance zone is located centrally, coaxially or symmetrically with respect to the theoretical exact (nominal) location (Figures 3.4–3.9).

With **orientation tolerances**, the tolerance zone is orientated with respect to the datum. The distance from the datum is not limited.

With **form tolerances**, the distance and orientation of the tolerance zone with respect to other features (form elements) are not limited.

Exceptions are tolerance zones of form tolerances where the nominal (geometrical ideal) form is defined by theoretical exact dimensions related to datums, (Figures 4.2 and 20.11). In these cases the tolerance zone is symmetrically placed (equidistant) with respect to the nominal form (line or face) in the nominal (theoretical exact) location and orientation with respect to the datum(s). The ways of defining the tolerance zone are imaginary circles or spheres. Their centres are located on the nominal form. The envelopes (tangential) on the circles or spheres form the tolerance zone (Figures 3.1 (top) and 20.10).

Unilaterally disposed tolerance zones (on one side of the nominal form only) are not provided with ISO 1101. In all cases the theoretical exact dimensions should be chosen in such a way that the tolerance zone limits are equidistant from the nominal form. This avoids the sometimes difficult estimation (in cases of curved forms) of the geometrical ideal form following the tolerance zone centres during manufacturing and inspection. According to ANSI Y14.5 M, the tolerance zone may be disposed unilaterally (see 2.1.1.1).

3.3.3 Width of the tolerance zone

Tolerance zones have limiting lines or faces that are equidistant from the nominal (geometrical ideal) form. The shape of the tolerance zone is independent of the size (diameter or distance) of the feature(s). Exceptions are tolerance zones of form of any line or surface where the nominal (geometrical ideal) form is defined by theoretical exact dimensions.

The width of the tolerance zone (in the measuring direction) is equal to the tolerance value, and is directed according to ISO 1101 in the direction of the leader line arrow:

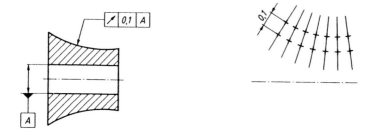

Figure 3.25: Width of the tolerance zone in the direction of the leader line arrow perpendicular to the surface.

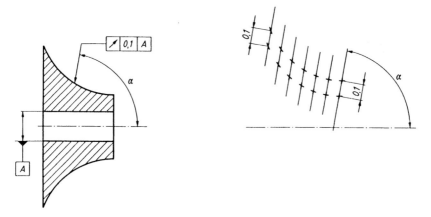

Figure 3.26: Indicated direction of the width of the tolerance zone.

- with tolerance of axes and median faces perpendicular to the toleranced axis or median face,

- with tolerances of surfaces or generator lines perpendicular to the surface (Figure 3.25).

An exception is the tolerance zone of roundness, where, if not otherwise specified, the width is defined perpendicular to the axis. (However, with run-out tolerances, if not otherwise specified, the width of the tolerance zone is defined perpendicular to the surface; Figure 3.25).

When, in exceptional cases (e.g. with cones or curves), the width of the tolerance zone is not perpendicular to the surface or the axis or the median face, the direction of the width of the tolerance zone should be indicated as shown in Figure 3.26.

3.3.4 Length of the tolerance zone

If not otherwise indicated, the geometrical tolerance (tolerance zone) applies to the entire length or surface of the feature.

When the leader line arrow points to a thick chain line or area within a thick chain line (Figure 3.27), the tolerance applies to this region. When the line designating the area within a thick chain line is dashed, the indication applies to the hidden surface (Fig. 3.27c).

When the tolerance value is followed by an oblique stroke and another value (e.g. 0.05/100), the second value (100 here) indicates the length within which the tolerance (0.05 here) applies. The tolerance zone (of 0.05 width) of the specified length (100) applies to all possible locations on the feature. Therefore these tolerances (0.05) can accumulate along the feature (Figure 3.28) if there is no

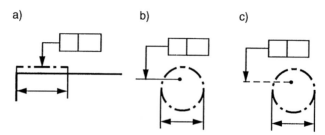

Figure 3.27: Tolerance zone applied to a restricted region: (a, b) on the surface shown; (c) on the hidden surface.

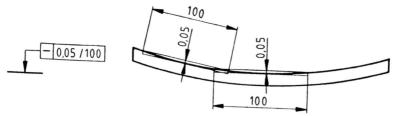

Figure 3.28: Tolerance zones applied to restricted lengths in any location.

further limitation by a tolerance of form, orientation, location or total run-out that the tolerance zone reaches over the entire feature.

3.3.5 Common tolerance zone

Isolated tolerance zones having the same value applied to different features may be indicated as shown in Figure 3.29.

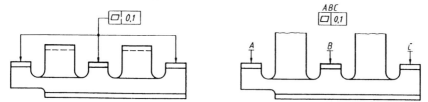

Figure 3.29: Simplified indication of tolerances that are similar to but independent of each other.

When the features are to be contained in a common tolerance zone, this should be indicated by the indication "common zone" above the tolerance frame (Figures 3.30 and 20.41).

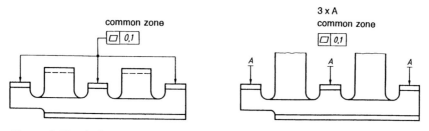

Figure 3.30: Indication of a common tolerance zone.

33

3.4 DATUMS

Datums define the orientation and/or location of the tolerance zone. For this purpose, the form deviations of the datum feature(s) are eliminated by the minimum-rock requirement (see 18.3.4) or by the indication of datum targets (Figure 3.36).

The datum may be established by

- one single datum feature;

- two or more datum features of same priority as a common datum (e.g. a common axis, Figure 3.31), indicated by a hyphen between the datum letters in the tolerance frame;

- two or more datum features with different but not specified priorities (indicated by a sequence of datum letters in the tolerance frame without separation) (Figure 3.32); application of this should be avoided because it is not unequivocal and may lead to different measuring results for the same workpiece (Figure 3.35); This indication will therefore be eliminated from ISO 1101 in the next edition;

- datum features of different priority (e.g. a three-plane datum system according to ISO 5459) (Figures 3.33–3.35)

- datum targets of different priority on different features (Figure 3.36)

Datums can have different priorities (order of precedence) as follows (Figures 3.34 and 3.35):

- **primary datum**, orientated according to the minimum rock-requirement (see 18.3.4) relative to the auxiliary datum element;

- **secondary datum**, orientated without tilting relative to the primary auxiliary datum element (only by translation and rotation) according to the minimum-rock requirement relative to the secondary auxiliary datum element (the secondary auxiliary datum element is perpendicular to the primary auxiliary datum element);

- **tertiary datum**, positioned without tilting and rotation relative to the primary and to the secondary auxiliary datum element only by translation until contact with the tertiary auxiliary datum element (the tertiary auxiliary datum element is perpendicular to the primary and secondary auxiliary datum elements).

The effect of different orders of priority is illustrated in Figure 3.35.

Form deviations of the datum features are not limited by the indication as a datum. Therefore, if the general form tolerance (see 16) is not sufficient, individual form tolerances must be indicated on the datum feature or datum targets according to ISO 5459 must be applied (Figure 3.36).

When datum features may have relatively large form deviations compared with the tolerance of orientation or location (e.g. with castings), **datum targets** according to ISO 5459 may be indicated (Figure 3.36).

A datum target may be a restricted area, a line or a point of specified location, see Table 3.1. They represent the supports of the workpiece during inspection (and manufacturing).

In indicating datum targets there should be specified:

- at the primary datum   3 targets
- at the secondary datum   2 targets
- at the tertiary datum   1 target

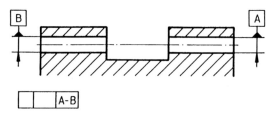

Figure 3.31: Datums of same priority as common datums.

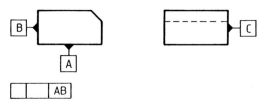

Figure 3.32: Datums with priorities not specified.

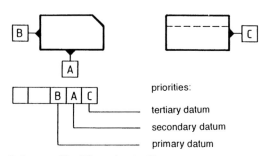

Figure 3.33: Datums with different priorities.

35

Principles of Geometrical Tolerancing

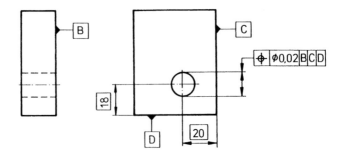

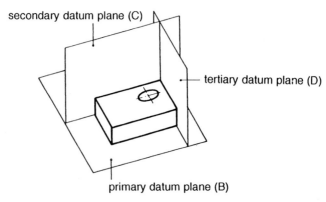

Figure 3.34: Three-plane datum system according to ISO 5459.

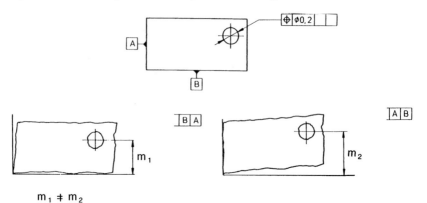

$m_1 \neq m_2$

Figure 3.35: Datums with different orders of priority.

Exceptions to the rules given above are datum features where the datum letter in the tolerance frame is followed by the symbols Ⓜ or Ⓛ (see 9.2 and 9.3.3; 11.2 and 11.3.3).

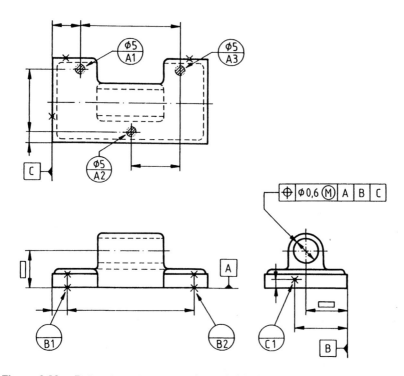

Figure 3.36: Datum targets: targets (areas) A1, A2, A3 establish the primary datum A; targets (lines) B1, B2 establish the secondary datum B; target (point) C1 establishes the tertiary datum C.

Datum features of too short length create great measuring uncertainty. Therefore these short features should not serve as datums. If necessary, the drawing must be altered (Figure 19.13).

3.5 AXES AND MEDIAN FACES

When the leader line arrow is aligned with the dimension line, the geometrical tolerance applies to the axis or median face, (Figure 3.37). The form deviations of the generator lines or surfaces are not limited, they may have larger straightness or flatness deviations. When the leader line arrow is indicated on the centre line in the drawing, the geometrical tolerance applies to all axes or median faces that the centre line represents. (The latter indication should be avoided because the standard (ISO 1101) does not specify whether a common tolerance zone is required or not.)

When the leader line arrow is indicated apart from the dimension line (the separation should be at least 4 mm), the geometrical tolerance applies directly to the surface, generator line or circumference line (Figure 3.38). This indication limits also the straightness deviation of the axis, since the axis cannot deviate more from straightness than the generator lines (Figure 3.39). A similar consideration applies to the flatness deviation of the median face.

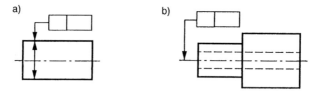

Figure 3.37: Tolerance indication for axes and median faces: (a) axis or median face; (b) all axes.

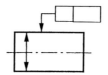

Figure 3.38: Tolerance indication for surfaces, generator lines and circumference lines.

A similar symbolization applies to datums. When the datum triangle and the dimension line are aligned, the axis or median face is the datum. When the datum triangle is indicated apart from the dimension line, the minimum-rock requirement applies to the surface or to the generator lines.

When the datum triangle is indicated on the centre line in the drawing, all axes or median faces that the centre line represents apply as a common datum. This is the axis of the smallest cylinder containing all axes or the median plane of two parallel planes of least distance containing all median faces. According to ISO

1101, this indication should be avoided because of the difficulties that arise with inspection.

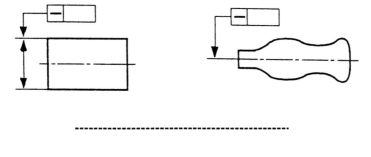

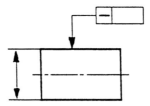

Figure 3.39: Straightness tolerance of axes and generator lines.

With axes it is necessary to distinguish between nominal axes, actual axes and datum axes.

Nominal axis This is the geometrical ideal (straight) axis of the geometrical ideal feature of nominal size defined in the drawing.

Actual axis This is the irregular axis that follows the form deviations of the feature. The definition of the actual axis is not yet standardized. The following definitions are under discussion. The axis is composed of the centres of

(a) least-squares circles (Gauss),
(b) minimum zone circles (Chebyshev),
(c) contacting circles,

in cross-sections perpendicular to the axis of the

(a) least-squares cylinder (Gauss),
(b) minimum-zone cylinders (Chebyshev),
(c) contacting cylinder.

The definitions of these circles and cylinders are given in 18.7.2.1 and 18.7.5.1. In the case of a cone the least-squares (substitute) cone is the best-fit cone, i.e. the (substitute) cone angle deviates from the nominal cone angle (see 8.3.1 and Figure 8.22).

The least-squares circles and cylinders or cones have advantages in measurement (least amount of necessary measured points and least measuring time). The least-squares circles are always unique (for all types of roundness deviations). This does not apply to the minimum-zone circles or to the contacting circles.

Minimum-zone circles and cylinders and cones have the advantage of being in harmony with long existing specifications in ISO 1101.

Contacting circles and cylinders and cones have the advantage of relating to those parts of the workpiece surface that are in contact with the counterpart in cases of fits. But they have the disadvantage of relating to just a few points of the surface, and therefore are rather unsuitable for assessing the dynamic behaviour (imbalance) of the workpiece.

In the past it was thought to be unnecessary to define the actual axis. But recent investigations reveal that considerable differences (up to 100%) in the measurement result can arise owing to these differences in definition. Therefore it is necessary, especially for the programming of coordinate measuring machines and form measuring machines, to standardize definitions unequivocally. An ISO Standard dealing with this subject is in preparation.

As this ISO Standard is not yet published, in cases of dispute it must be agreed upon whether the workpiece is to be accepted when it meets the drawing specifications:

- using one of the definitions given above;
- using one specific of the definitions given above.

The author recommends the use of the least-squares method for the definition of an axis, because

- it is the only method that is always unique;

- all assessed points of the workpiece contribute to the determination of the axis (in cases of fits and of material thickness the boundary (E), (M), (L) is of interest rather than the axis);

- it needs the least number of points to become sufficiently stable (to meet the required measurement uncertainty).

See 8.3.1 and Figure 8.26.

Datum axis   This is the axis of the contacting cylinder or cone orientated according to the minimum-rock requirement, ISO 5459 (see 18.3.4). Again here it is under discussion whether the definition of ISO 5459 should be changed to the least-squares cylinder or cone (at present coordinate measuring machines use this definition).

Similar considerations apply to median planes, median faces and the datum median plane.

Median plane **(nominal median plane)**   This corresponds to the nominal axis, and is the symmetry plane of the geometrical ideal feature of nominal size defined in the drawing.

Median face **(actual median face)**   This corresponds to the actual axis; for example, in the case of the least-squares method the median face is composed of the centre points between opposite points on opposite surfaces, while the connection of each pair of points is perpendicular to the least-squares substitute median plane (the median plane between the substitute planes of the two opposite surfaces) (see 8.3.1 and Figure 8.25).

Datum median plane   This corresponds to the datum axis, and is the median (symmetry) plane of the contacting pair of parallel planes orientated according to the minimum-rock requirement, ISO 5459. Again here it is under discussion whether the definition of ISO 5459 should be changed to the symmetry plane of the pair of least squares substitute planes.

Candidate datum method   At present discussions are under way in the ISO and the ANSI regarding whether the minimum-rock requirement should be substituted by a candidate datum method (see 18.3.4) rather than the least-squares method.

3.6   SCREW THREADS, GEARS AND SPLINES

Tolerances of orientation or location and datum references specified (indicated) for screw threads apply to the axis of the thread derived form the pitch cylinder, if not otherwise specified (Figures 3.40 and 3.41).

Tolerances of orientation or location and datum references specified (indicated) for gears and splines must designate the specific feature of the gear or spline to which they apply (e.g. pitch diameter PD, major diameter MD or minor (least) diameter LD).

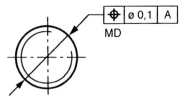

Figure 3.40: Positional tolerance specified for the major diameter cylinder axis of a thread.

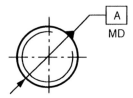

Figure 3.41: Datum specified as the major diameter cylinder axis of a thread.

3.7 ANGULARITY TOLERANCES AND ANGULAR DIMENSION TOLERANCES

Angular dimension tolerances are usually specified in angular grades or minutes following the nominal angular size (e.g. $45° \pm 30'$). Angular dimension tolerances define permissible ranges of angles (e.g. $44° 30'–45'° 30'$) for the general direction of actual lines or line elements of surfaces (contacting lines).

The general direction of the contacting line derived from the actual surface is the direction of the contacting line of geometric ideal form (straight line). The maximum distance of the actual line from the contacting line is the least possible, i.e. the contacting line (e.g. leg of the angle measuring instrument) contacts the actual surface on the highest point(s) in an average direction (Figure 3.42). Therefore the angular dimension tolerance does not limit the form deviations of the lines or surfaces constituting the angle.

Angularity (inclination) tolerances, specified by the symbol $<$ in the tolerance frame and by a tolerance value in mm (e.g. $< \boxed{0.2 \, \text{A}}$) define tolerance zones within which all points of the toleranced surface or line must be contained (Figure 3.43). Therefore the inclination tolerance limits also the form deviations of the toleranced feature.

42

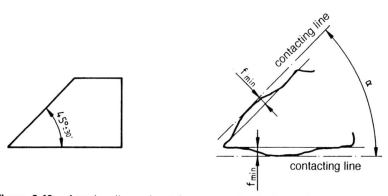

Figure 3.42: Angular dimension tolerance and actual angular size.

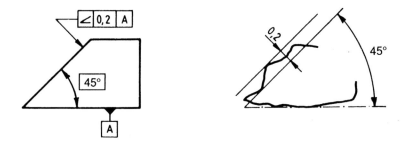

Figure 3.43: Angularity (inclination) tolerance and tolerance zone.

3.8 TWIST TOLERANCE

Some national standards define a twist tolerance instead of the flatness tolerance (e.g. the German Standards on aluminium sections DIN 1748 and DIN 176215). The definition is as follows. The section is laid on a flat ground. One end should contact the ground as perfectly as possible (minimum-rock requirement). At the other end the twist deviation is measured as the maximum distance from the ground (Figure 3.44). The twist tolerance is the maximum permissible value of the twist deviation.

At the same piece of section the twist deviation δ_v is larger than the flatness deviation δ_e because for the assessment of the flatness deviation the entire surface of the section (not only one end) must follow the minimum requirement (relative to the flat ground) (Figure 3.45).

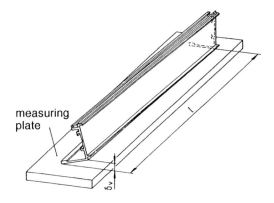

measuring plate

Figure 3.44: Twist deviation of a section.

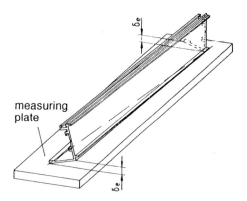

measuring plate

Figure 3.45: Flatness deviation of a section.

4 Profile Tolerancing

Profile tolerances are defined in ISO 1101, and profile tolerancing is described in ISO 1660.

With profile tolerancing a distinction must be made between tolerancing of the form of lines (symbol ⌒) and of the form of surfaces (symbol ⌒).

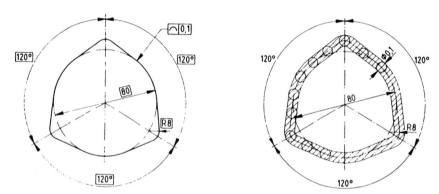

Figure 4.1: Profile tolerance of lines; the nominal form is defined by consecutive straight lines and circles defined by theoretical exact dimensions without reference to datums.

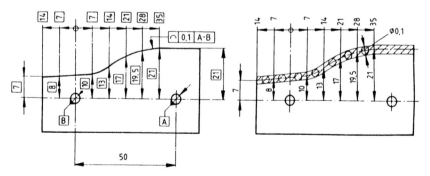

Figure 4.2: Profile tolerance of lines; the nominal form is defined by theoretical exact dimensions with reference to datums (A–B).

Profile Tolerancing

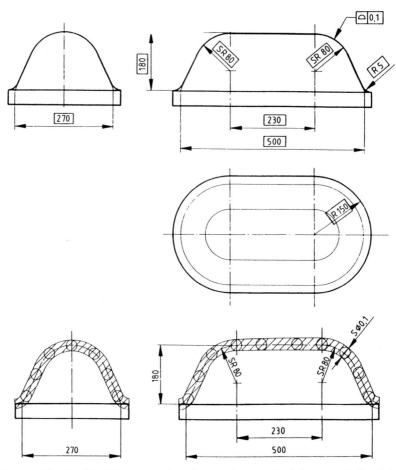

Figure 4.3: Profile tolerance of surfaces; the nominal form is defined by consecutive theoretical exact planes, spheres, cylinders and tori, which are defined by theoretical exact dimensions without reference to datums

The nominal (theoretical exact, geometrical ideal, true) form is to be defined by theoretical exact (rectangular framed) dimensions with (Figures 4.2 and 20.12) or without (Figures 4.1, 4.3 and 20.10) relation to datum(s).*

---

*With plain geometrical forms (straightness, roundness, flatness and cylindricity), the drawing defines the nominal form without theoretical exact dimensions. Then, however, the use of the special symbols –, ○, ◻, ⋈, is recommended.

46

The tolerance zone of a profile tolerance is defined by (tangential) envelopes on circles (profile tolerance of a line) or spheres (profile tolerance of a surface) whose diameters are equal to the tolerance value and that are taken to be centred on the nominal form (Figures 4.1–4.3). Therefore the zone is equally disposed on either side of the nominal profile.

Former practice allowed disposition of the whole tolerance on one side of the nominal profile (within the material, −, or outside it, +). According to ISO 1101, this is not recommended because it creates difficulties in the proper definition of the tolerance zone when there are edges or narrow or changing curvatures within the profile. If necessary, the nominal dimensions should be recalculated (and the drawing changed accordingly) in order to meet the requirements of the profile tolerance according to ISO 1101 and ISO 1660 (symmetrically disposed tolerance zone).

When the width of the tolerance zone varies within the profile (see Figure 20.10), the form of the envelope of the circles or spheres (tolerance zone) between the specified points (e.g. splines) is not standardized. Clearly, the envelope will alter its form in a smooth manner.

5 Tolerancing of Cones

5.1 GENERAL

With cones, the following characteristics may be toleranced:

- form of cone (conicity);
- orientation and radial location of the cone (cone axis) relative to a datum;
- axial location of the cone relative to a datum;
- distances of the endfaces of the truncated cone relative to a datum.

The first three characteristics (deviations) may be toleranced either together (in common) by the form tolerance of the cone (Figure 5.12) or separately by different tolerances, (Figures 5.3–5.11).

The cone angle may be specified as the theoretical exact angular dimension or as the theoretical exact cone ratio (Figures 5.1 and 5.2).

In order to define the axial location of the cone, the following may be specified:

- the theoretical exact cone diameter and theoretical exact distance of this diameter from a datum (Figure 5.3) when the axial deviation is limited by the form tolerance of the cone (Figure 5.13), T_x;
- the theoretical exact cone diameter and nominal distance of this diameter from a datum and permissible deviations of the distance (Figure 5.4) when the axial deviation is larger than that according to the form tolerance of the cone (Figure 5.13), T_x;
- the theoretical exact gauge dimension (maximum material size) with nominal distance from a datum and permissible deviations of this distance (Figure 5.8).

In order to define the radial location of the cone (cone axis) relative to a datum axis, the following may be specified:

- the form tolerance of the cone related to a datum axis (Figure 5.9), when the radial deviation is limited by the form tolerance of the cone;
- the coaxiality tolerance of the cone axis relative to the datum axis (Figure 5.10) when the radial deviation is larger than that according to the form tolerance of the cone;

• the run-out tolerance of the cone (instead of the coaxiality tolerance) relative to the datum axis (Figure 5.11) when the roundness deviations is included (the run-out deviation is easier to inspect than the coaxiality deviation).

The following figures show on the left the drawing indications and on the right the tolerance zones of the cone.

5.2 FORM TOLERANCE AND DIMENSIONING OF THE CONE

For dimensioning and tolerancing the form of cones see Figures 5.1 and 5.2.

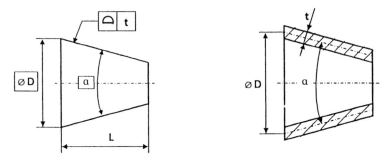

Figure 5.1: Cone tolerancing by form tolerance, theoretical exact maximum cone diameter and theoretical exact cone angle.

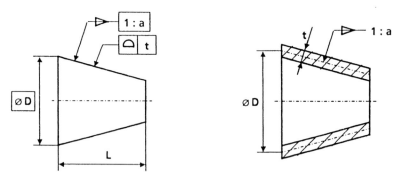

Figure 5.2: Cone tolerancing by form tolerance, theoretical exact maximum cone diameter and theoretical exact cone ratio.

Instead of the theoretical exact maximum cone diameter, another theoretical exact cone diameter may be specified (Figures 5.3, 5.4 and 5.12).

The definition of the theoretical exact cone by the specification of the theoretical exact maximum and minimum cone diameters and the theoretical exact distance between them is not recommended, since it would necessitate the indication in addition of the same distance as the nominal dimension with permissible deviations, and might therefore cause confusion.

5.3 TOLERANCING OF THE AXIAL LOCATION OF THE CONE

Usually cones and truncated cones are not the only features of a workpiece. Then the axial location of a specified diameter of the cone relative to a datum should be dimensioned and toleranced (Figures 5.3 and 5.4).

According to Figure 5.3, the axial location of the cone is limited by the form tolerance t (see 5.6). When the location of the cone may vary more than in Figure 5.3 larger tolerances (see $\pm$ $T/2$ in Figure 5.4) may be applied.

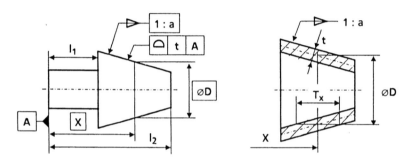

Figure 5.3: Cone tolerancing by form tolerance t, theoretical exact cone diameter D, theoretical exact distance X of this cone diameter D from the datum A, and theoretical exact cone ratio 1:a (or cone angle). The axial location of the cone ($\varnothing$ D) is limited between $X \pm T_x/2$, and the axis of the form tolerance zone is perpendicular to the datum A.

When with dimensioning according to Figure 5.3 X and D coincide on the drawing with a cone end face (Figure 5.5), the limits of the cone end face diameter are defined by the form tolerance and the theoretical exact cone end face diameter. These limit dimensions and the form tolerance define the limits of the distance of the cone end face from the datum A. The theoretical exact dimension X serves only to define the location of the form tolerance zone. Therefore X serves only indirectly for the definition of the location (distance from datum A) of the cone end face. (The tolerance of the distance X between the two left end faces is non-zero. A distinction must be made between the theoretical exact dimension X, which serves to define the location of the cone tolerance zone, and the permissible range of the distance X of the actual cone end face from the datum A, which is defined by the result of the cone tolerancing.)

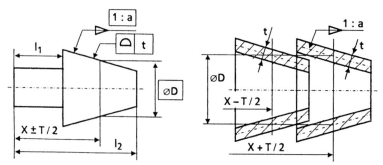

Figure 5.4: Cone tolerancing by form tolerance t, theoretical exact cone diameter D, dimension X and dimensional tolerance $\pm T/2$ to limit the location of this cone diameter D, theoretical exact cone ratio $1 : a$ (or cone angle). The axial location of the actual cone diameter D is limited by $X \pm (T/2 + T_x/2)$ (see 5.6).

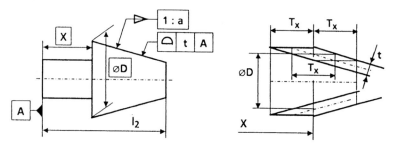

Figure 5.5: Cone tolerancing by form tolerance t, theoretical exact maximum cone diameter (theoretical exact end face diameter) D, theoretical exact distance X of this cone diameter from a datum, and theoretically exact cone ratio $1 : a$ (or cone angle). The axial location of the actual cone diameter D is limited by $X \pm T_x/2$; the cone end face is $X \pm T_x$ from the datum A apart; and the axis of the form tolerance zone is perpendicular to the datum A.

When the distance X between the cone end face and the end face of the part may vary by more than $\pm T_x$ (Figure 5.5), tolerancing according to Figures 5.6 and 5.7 may be applied.

When dimensioned according to Figure 5.6 or 5.7, i.e. when X and D are indicated at a cone end face, the tolerance $\pm T/2$ of X applies to both (a) the location of the theoretically exact cone diameter D of the form tolerance zone and (b) the location of the cone end face. The difference between the methods of Figures 5.6 and 5.7 is that, according to the latter the axis of the form tolerance

5.3 Tolerancing of the axial location of the cone

zone is perpendicular to the cone end face (datum A), but this is not necessarily so according to Figure 5.6.

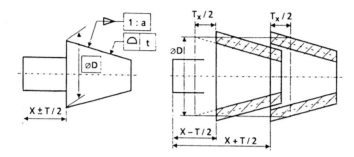

Figure 5.6: Cone tolerancing by form tolerance t, theoretical exact cone diameter D (here the maximum nominal cone diameter), dimension X and dimensional tolerance $\pm T/2$ to limit the location of this cone diameter D and also to limit the location of the cone end face, and theoretical exact cone ratio $1:a$, (or cone angle). The axial location of the actual cone diameter D (existing or imagined as an extension of the actual cone) is limited by $X \pm (T/2 + T_x/2)$ (see 5.6), and the axial location of the cone end face is limited by $X \pm T/2$.

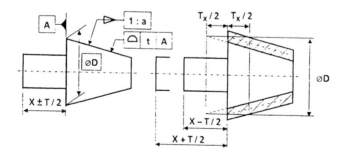

Figure 5.7: Cone tolerancing by form tolerance t, theoretical exact cone diameter D, dimension X and dimensional tolerance $\pm T/2$ to limit the location of this cone diameter D, theoretical exact cone ratio $1:a$ (or cone angle). The axial location (distance) of the actual cone diameter D (existing or imagined as an extension of the actual cone) relative to the parts left end face is limited by $X \pm (T/2 + T_x/2)$ (see 5.6), the actual location of the part end face relative to the datum A is limited by $X \pm T/2$, and the axis of the form tolerance zone is perpendicular to the datum A.

53

A frequently used method of tolerancing the axial cone location is shown in Figure 5.8. It has the advantage of easy inspection when an appropriate gauge can be used.

Tolerancing of the actual location of the cone as shown in Figures 5.3–5.8 does not limit the deviations of orientation and radial location of the cone. In order to limit all aspects, a combination of the methods shown in Figures 5.3–5.8 and in Figures 5.9–5.11 or tolerancing according to Figure 5.12 should be applied.

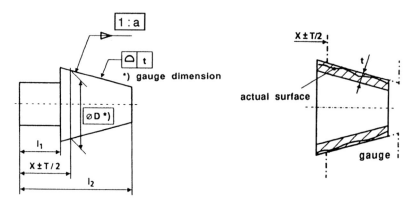

Figure 5.8: Cone tolerancing by form tolerance t, theoretical exact gauge dimension $\varnothing\,D$,* dimensional tolerance $\pm\,T/2$ to limit the actual location of the theoretical exact gauge dimension $\varnothing\,D^*$, and theoretical exact cone ratio $1:a$ (or cone angle)

5.4 TOLERANCING OF THE ORIENTATION AND RADIAL LOCATION OF THE CONE

The orientation and radial location of the cone can be toleranced as shown in Figures 5.9–5.11. The procedures shown in Figures 5.10 and 5.11 allow the specification of wider or narrower tolerances of the orientation and radial location of the cone than that according to Figure 5.9.

The tolerancings of the orientation and radial location of the cone as shown in Figures 5.9–5.11 do not limit the deviation of the axial location of the cone. In order to limit all aspects, a combination of the methods shown in Figures 5.3–5.8 and in Figures 5.9–5.11 or tolerancing according to Figure 5.12 should be applied.

5.5 Related form tolerance for tolerancing form, orientation and radial and axial location of the cone

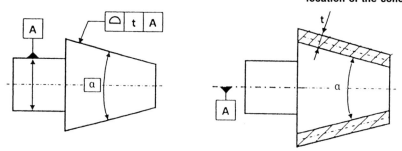

Figure 5.9: Cone tolerancing by form tolerance *t*, related to a datum axis A, and theoretical exact cone angle α (or cone ratio). For the sake of clarity, the necessary indications defining the cone diameter and its axial location have been omitted in this drawing.

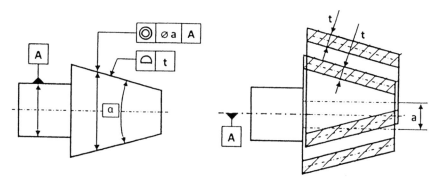

Figure 5.10: Cone tolerancing by form tolerance *t*, coaxiality tolerance ⌀ a for the orientation and radial location of the cone and theoretical exact cone angle α (or cone ratio). For the sake of clarity, the necessary indications defining the cone diameter and its axial location have been omitted in this drawing. The cone diameter at the leader line arrow of the coaxiality tolerance has no specified dimension in order to indicate that the entire length of the truncated cone is toleranced. The actual axis of the truncated cone must be contained in the coaxiality tolerance zone ⌀ a.

5.5 RELATED FORM TOLERANCE FOR TOLERANCING FORM, ORIENTATION AND RADIAL AND AXIAL LOCATION OF THE CONE

Figure 5.12 shows tolerancing of a cone when all aspects (form, orientation, radial and axial location of the cone) are limited by the form tolerance *t*.

55

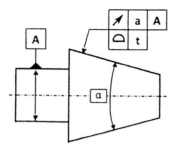

Figure 5.11: Cone tolerancing by form tolerance t, run-out tolerance a for the orientation and radial location of the cone, and theoretical exact cone angle (or cone ratio). For the sake of clarity, the necessary indications defining the cone diameter and its axial location have been omitted in this drawing.

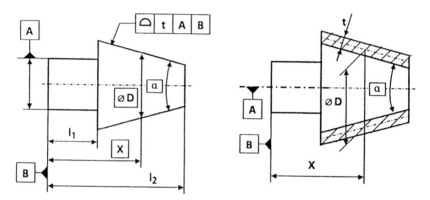

Figure 5.12: Cone tolerancing by form tolerance t related to a datum system (limits form, orientation, radial and axial location).

5.6 RELATIONSHIP BETWEEN THE CONE TOLERANCES

The value of the form tolerance t of a cone applies in the direction normal to the cone surface (diameter t of imaginary spheres centred at the theoretical exact cone). The form tolerance can be converted as follows:

- tolerance T_D normal to the cone axis (Figure 5.13)

$$T_D/2 = t/\cos(\alpha/2), \qquad t = (T_D/2)\cos(\alpha/2);$$

- axial cone tolerance T_x (permissible location of a theoretically exact cone diameter) (Figure 5.13)

$$T_x = t/\sin(\alpha/2), \qquad t = T_x \sin(\alpha/2);$$

- cone angle (Figure 5.14)

$$\alpha_{max} = 2 \arctan\frac{l \tan(\alpha/2) + T_D/2}{l}$$

$$\alpha_{min} = 2 \arctan\frac{l \tan(\alpha/2) - T_D/2}{l}$$

$$t = l \cos(\alpha/2)\,\{\tan(\alpha/2) - \tan[(\alpha/2) - (T_\alpha/4)]\}$$

When the cone ratio is $1:3$ or smaller, or when the cone angle is 20° or smaller, the difference between t and $T_D/2$ is less than 2% and therefore practically negligible.

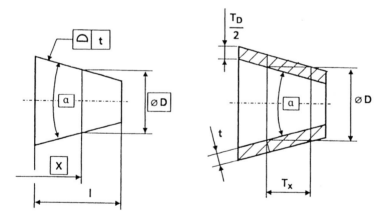

Figure 5.13: Conversion of cone form tolerance t, cone diameter tolerance T_D and tolerance of axial cone location T_x.

ISO 1947 still uses cone angle tolerances T_α and cone diameter tolerances T_D. ISO 5166 still uses axial cone tolerances T_x. These tolerances can be converted into cone form tolerances t as above.

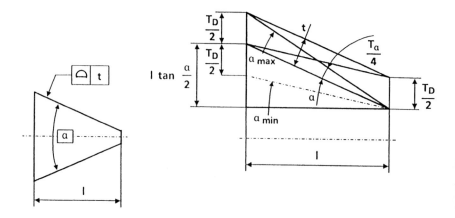

Figure 5.14: Conversion of cone form tolerance t, cone angle tolerance $\pm T_\alpha/2$ or cone angle limit values α_{min} and α_{max}.

6 Positional Tolerancing

6.1 DEFINITION

The positional tolerance is defined in ISO1101, and positional tolerancing is described in ISO 5458.

In the method of positional tolerancing (for the location of features) theoretical exact dimensions and positional tolerances determine the location of features (points, axes, median faces and plane surfaces) relative to each other or in relation to one or more datum(s). The tolerance zone is symmetrically disposed about the theoretical exact location.

By virtue of this definition, positional tolerances do not accumulate where theoretically exact dimensions are arranged in a chain (Figures 6.1, 6.2 and 18.55). This contrasts with dimensional tolerances arranged in a chain.

6.2 THEORETICAL EXACT DIMENSIONS

According to ISO 1101, theoretical exact dimensions are to be indicated in rectangular frames.

However, the following theoretical exact dimensions that determine the theoretical exact locations of positional tolerance zones are not to be indicated:

(a) theoretical exact angles between features (e.g. holes) equally spaced on a complete pitch circle (Figures 6.4 and 6.5); and

(b) theoretical exact dimensions 90° and 0°, 180° or distance 0 between

- positional-toleranced features unrelated to a datum (Figures 6.2, 20.19 and 20.70),

- positional-toleranced features related to the same datum(s) (Figures 3.24 and 6.4),

- positional-toleranced features and their related datum(s) (Figure 6.1).

When the positional toleranced features are drawn on the same centre line, they are regarded as related features having the same theoretically exact location (Figure 6.4), unless otherwise specified, for example by relation to different datums (Figure 3.23) or by an appropriate note on the drawing (Figure 6.5). In Figure 6.5 the locations of the two patterns of features (holes) are independent of each other. See also 20.6.5.

59

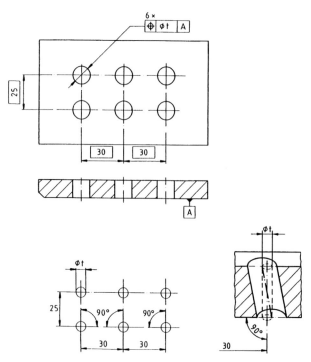

Figure 6.1: Positional tolerancing.

However datums are not necessarily perpendicular to each other when they are not in the same datum system. (This applies to all datums of any geometrical tolerancing.)

6.3   FORM OF THE POSITIONAL TOLERANCE ZONE

The width of the tolerance zone has the direction of the leader line arrow (which connects the tolerance feature with the tolerance frame) (Figure 6.2).

With the drawing indication according to Figure 6.2 but without the "axis" indications, the median faces of the two opposite plane surfaces would be tolerances. The tolerance zone would then be the space between two parallel planes of distance 0.3 along the entire width of the hole. However, in ISO 5458:1987 for these examples the tolerance zone is referred to as being applicable to the axes of the (rectangular) holes. Therefore, in order to avoid misunderstanding, it is recommended that in the case, of rectangular features the indication "axis" should be added to the tolerance frame when the axes are toleranced (Figure 6.2),

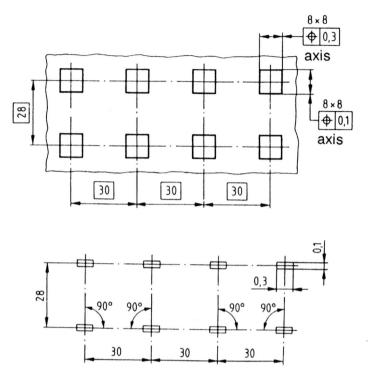

Figure 6.2: Positional tolerance zones with perpendicular cross-sections.

or the indication "median faces" when the median faces are toleranced.

When the tolerance value is preceded by the symbol $\emptyset$ the tolerance zone is cylindrical.

For cylindrical features of mating parts, the tolerance zone is usually cylindrical, because

- the function permits the same deviation in all directions (multidirectional) from the theoretically exact location;

- manufacturing causes multidirectional deviations.

Positional tolerancing provides an indication of cylindrical tolerance zones. This contrasts with dimensional tolerancing, which can only generate rectangular tolerance zones. Cylindrical tolerance zones are 57% larger than the corresponding square tolerance zones (Figure 6.3).

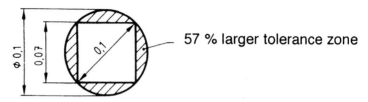

57 % larger tolerance zone

Figure 6.3: Comparison of tolerance zones with round and square cross-section.

6.4 POSITIONAL TOLERANCES ON A CIRCLE

With tolerancing of hole distances on a pitch diameter, positional tolerancing provides the advantage of specifying the theoretical exact location of the feature. This is a prerequisite for proper tolerance calculation and proper inspection.

Dimensional tolerancing, however, results in tolerance accumulation and, on a closed pitch circle, in contradictory tolerances. Therefore proper tolerance calculations and inspections are not possible with dimensional tolerancing on a closed pitch circle.

6.5 POSITIONAL TOLERANCES RELATED TO A DATUM

Positional tolerances may be related to one or more datum(s) in an unequivocal way (Figures 6.1 and 6.8). Where positional tolerance zones are perpendicular to the related datum, the theoretical exact angle 90° need not to be indicated in the drawing (see 6.2 and Fig. 6.1). Where this relationship is omitted (Figure 6.1 without datum indication), the positional tolerance zones are parallel to each other, but they need not to be perpendicular to the faces.

6.6 TOLERANCE COMBINATIONS

6.6.1 Combination of dimensional coordinate tolerancing and positional tolerancing

When features within a group are individually located by positional tolerancing and their pattern is located by coordinate dimensioning and tolerancing, each requirement should be met independently (Figure 6.6) (ISO 5458).

For example the distances between the actual axis of each left hole and the left edge should be between the two size limits, 15.3 and 19.7.

The actual axis of each hole should be within the cylindrical positional tolerance

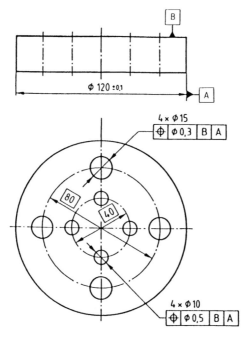

Figure 6.4: Positional-toleranced features (holes) equally spaced on a complete pitch circle; they are drawn on the same centre line, and are therefore regarded as related features having the same theoretical exact location.

zone ∅ 0.1. The positional tolerance zones are located in their theoretical exact locations relative to each other.

According to ANSI Y14.5 - 1973 and BS 308, this drawing indication has a different meaning. The dimensional coordinate tolerances apply to the location of the theoretical exact pattern of positional tolerances. Therefore, according to these standards, the permissible distance of the actual axis from the edge is half of the positional tolerance larger than according to ISO 5458 (Figures 6.7 and 21.2).

Because of the risk of misunderstanding, this method (dimensional coordinate tolerances adjacent to positional tolerances) should be avoided, and the method of combination of positional tolerances should be applied (see 6.6.2).

6.6.2 Combination of positional tolerances (composite positional tolerancing)

When features within a group are individually located by positional tolerancing and their pattern is located by different positional tolerancing, each requirement should be met independently (Figure 6.8) (ISO 5458).

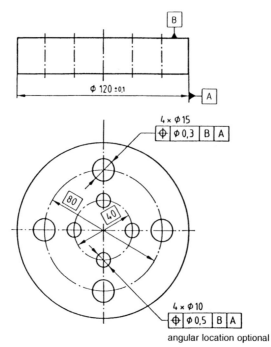

Figure 6.5: Positional-toleranced features (holes) drawn on the same centre line, but with angular location of the two patterns relative to each other being optional.

For example, the actual axis of each of the four holes should be within the cylindrical positional tolerance zone of $\varnothing$ 0.1. The positional tolerance zones are located in their theoretical exact positions in relation to each other and perpendicular to the datum A.

The actual axis of each hole should be within the cylindrical positional tolerance zone of $\varnothing$ 0.6. The positional tolerance zones are located in their theoretical exact positions in relation to the datums A, B and C.

6.7 CALCULATION OF POSITIONAL TOLERANCES

A distinction must be made between floating and fixed fasteners (Figures 6.9 and 6.10).

For **floating** fasteners (Figure 6.9) the positional tolerance t_{cd} of the holes is

$$t_{cd} = Ma_i - Ma_e = D_{min} - d_{max},$$

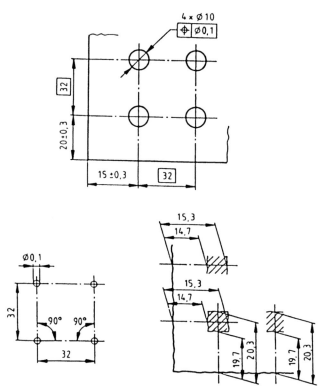

Figure 6.6: Dimensional coordinate tolerances adjacent to positional tolerances according to ISO 5458.

where Ma_i is the maximum material size of the internal dimension (e.g. the minimum size of the hole) and Ma_e is the maximum material size of external dimension (e.g. the maximum size of the bolt).

Figure 6.9 shows the extreme locations of the holes of maximum material size that still allow assembly with the bolt of maximum material size. From these locations, the positional tolerance zone $\varnothing$ ($D_{min} - d_{max}$) is derived.

For **fixed** fasteners (Figure 6.10) the positional tolerance t_{cc} of the holes is

$$t_{cc} = (Ma_i - Ma_e)/2 = (D_{min} - d_{max})/2.$$

Figure 6.10 shows the extreme locations of the holes of maximum material size that still allow assembly with the bolt of maximum material size. From these locations, the positional tolerance zone $\varnothing$ ($D_{min} - d_{max}$)/2 is derived.

The formulae are also applicable to non-cylindrical features (e.g. keys and key ways).

65

Positional Tolerancing

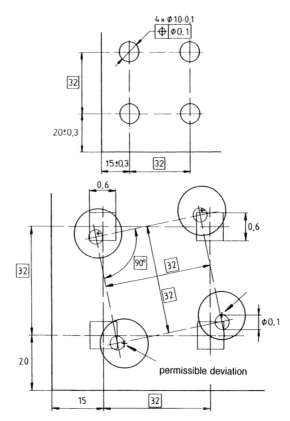

Figure 6.7: Dimensional coordinate tolerances adjacent to positional tolerances according to ANSI Y14.5 - 1973 and BS 308.

The indicated positional tolerances t_{ce} and t_{cd} do not take account of straightness tolerances of the screw and hole, the minimum clearance between bolt and hole, or any measurement uncertainty. If necessary, the values should be decreased accordingly. For head screws the under-head fillet may require chamfered holes or washers.

When the parts to be assembled (e.g. flanges) are centred without clearance (e.g. by a centring pin), the method for calculating the positional tolerance depends on the type of dimensioning.

In Figure 6.11 the centring pin hole is the datum (C) for the positional tolerance. (Manufacturing and measuring should start at the centering pin hole.) In this case the positional tolerance for the other holes is to be calculated as shown above.

66

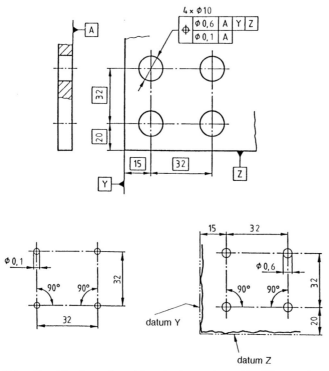

Figure 6.8:   Combination of positional tolerancing with different datums.

In Figure 6.12 for the centring pin hole a positional tolerance applies in relation to the other holes. (The centring pin hole is not necessarily the starting point for manufacturing.) In this case the positional tolerance of the other holes is to be reduced by the positional tolerance of the centring pin hole, i.e. for floating fasteners

$$t_1 + t_2 = D_{min} - d_{max},$$

while for fixed fasteners

$$t_1 + t_2 = (D_{min} = d_{max})/2.$$

The reason for the reduction is that the flanges may be fixed by the centring pin half of the centring pin hole positional tolerance, apart from the theoretical exact position relative to the other holes.

67

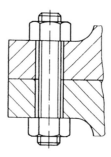

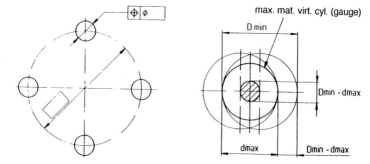

Figure 6.9: Positional tolerance with floating fastener.

6.8 ADVANTAGES OF POSITIONAL TOLERANCING

Positional tolerancing has the following advantages compared with dimensional coordinate tolerancing.

- Functional relationships are better indicated. Relationships to one or more datum(s) can be indicated unequivocally. Therefore function-related tolerancing with the largest possible tolerances is possible (see 20.6.4).

- There is the possibility of indicating cylindrical tolerances. Thereby 57% larger tolerances are possible compared with tolerances of rectangular cross-section (Figure 6.3). In many cases this corresponds to the function-related tolerancing (e.g. mating of cylindrical bolts with cylindrical holes).

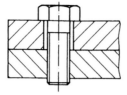

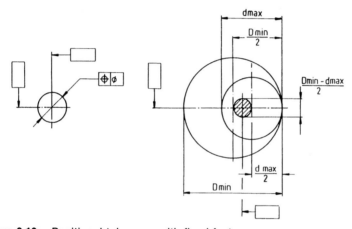

Figure 6.10: Positional tolerance with fixed fastener.

- There is a simple application of the maximum material requirement with additional gain of tolerance.

- There is the possibility of application of the projected tolerance zone method (see 7). Thereby function-related tolerancing (with the largest possible tolerances) becomes possible.

- There is no tolerance accumulation when theoretical exact dimensions are arranged in a chain. This enables, simple tolerance calculations. In some cases (e.g. holes on a pitch diameter) in practical positional tolerances only allow tolerance calculations and function-related, manufacturing-related and inspection-related tolerancing (see 20.6.4).

Positional Tolerancing

Table 6.1: Positional tolerances for holes according to ISO 273 and screws; t_{ce} is the positional tolerance for through-hole and threaded hole for a stud screw or head screw (Figure 6.10); t_{cd} is the positional tolerance for through-hole and a threaded bolt with nuts at both ends or a head screw with nut (Figure 6.9).

Dimensions in mm

Thread d	through-hole d_h								
	Fine	t_{ce}	t_{cd}	Medium	t_{ce}	t_{cd}	Coarse	t_{ce}	t_{cd}
1	1,1	0,05	0,1	1,2	0,1	0,2	1,3	0,15	0,3
1,2	1,3	0,05	0,1	1,4	0,1	0,2	1,5	0,15	0,3
1,4	1,5	0,05	0,1	1,6	0,1	0,2	1,8	0,2	0,4
1,6	1,7	0,05	0,1	1,8	0,1	0,2	2	0,2	0,4
1,8	2	0,1	0,2	2,1	0,15	0,3	2,2	0,2	0,4
2	2,2	0,1	0,2	2,4	0,2	0,4	2,6	0,3	0,6
2,5	2,7	0,1	0,2	2,9	0,2	0,4	3,1	0,3	0,6
3	3,2	0,1	0,2	3,4	0,2	0,4	3,6	0,3	0,6
3,5	3,7	0,1	0,2	3,9	0,2	0,4	4,2	0,35	0,7
4	4,3	0,15	0,3	4,5	0,25	0,5	4,8	0,4	0,8
4,5	4,8	0,15	0,3	5	0,25	0,5	5,3	0,4	0,8
5	5,3	0,15	0,3	5,5	0,25	0,5	5,8	0,4	0,8
6	6,4	0,2	0,4	6,6	0,3	0,6	7	0,5	1
7	7,4	0,2	0,4	7,6	0,3	0,6	8	0,5	1
8	8,4	0,2	0,4	9	0,5	1	10	1	2
10	10,5	0,25	0,5	11	0,5	1	12	1	2
12	13	0,5	1	13,5	0,75	1,5	14,5	1,25	2,5
14	15	0,5	1	15,5	0,75	1,5	16,5	1,25	2,5
16	17	0,5	1	17,5	0,75	1,5	18,5	1,25	2,5
18	19	0,5	1	20	1	2	21	1,5	3
20	21	0,5	1	22	1	2	24	2	4
22	23	0,5	1	24	1	2	26	2	4
24	25	0,5	1	26	1	2	28	2	4
27	28	0,5	1	30	1,5	3	32	2,5	5
30	31	0,5	1	33	1,5	3	35	2,5	5
33	34	0,5	1	36	1,5	3	38	2,5	5
36	37	0,5	1	39	1,5	3	42	3	6
39	40	0,5	1	42	1,5	3	45	3	6
42	43	0,5	1	45	1,5	3	48	3	6
45	46	0,5	1	48	1,5	3	52	3,5	7
48	50	1	2	52	2	4	56	4	8
52	54	1	2	56	2	4	62	5	10
56	58	1	2	62	3	6	66	5	10
60	62	1	2	66	3	6	70	5	10
64	66	1	2	70	3	6	74	5	10
68	70	1	2	74	3	6	78	5	10
72	74	1	2	78	3	6	82	5	10
76	78	1	2	82	3	6	86	5	10
80	82	1	2	86	3	6	91	5,5	11
85	87	1	2	91	3	6	96	5,5	11
90	93	1,5	3	96	3	6	101	5,5	11
95	98	1,5	3	101	3	6	107	6	12
100	104	2	4	107	3,5	7	112	6	12
105	109	2	4	112	3,5	7	117	6	12
110	114	2	4	117	3,5	7	122	6	12
115	119	2	4	122	3,5	7	127	6	12
120	124	2	4	127	3,5	7	132	6	12
125	129	2	4	132	3,5	7	137	6	12
130	134	2	4	137	3,5	7	144	7	14
140	144	2	4	147	3,5	7	155	7,5	15
150	155	2,5	5	158	4	8	165	7,5	15

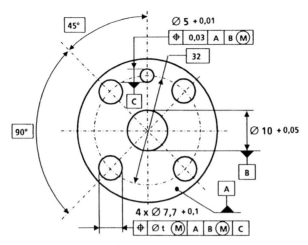

Figure 6.11: Positional tolerancing with centring pin hole as a datum. Through hole: $t = D_{min} - d_{max}$. Threaded hole + through hole: $t = (D_{min} - d_{max})/2$.

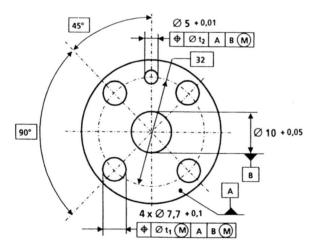

Figure 6.12: Positional tolerancing; centring pin hole, no datum. Through hole: $t_1 + t_2 = D_{min} - d_{max}$. Threaded hole + through hole: $t_1 + t_2 = (D_{min} - d_{max})/2$.

7 Projected Tolerance Zone

Positional tolerances of threaded holes and through-holes for fixed fasteners (Figures 6.10 and 7.1) are usually calculated from the maximum diameter of the bolt and the minimum diameter of the through-hole (see 6.7). A prerequisite for these calculations is that the axis is perpendicular to the joint surface. In the extreme case (Figure 7.2) the parts still fit. When the axis of the threaded hole deviates from the perpendicular orientation (but remains in the positional tolerance zone), the parts do not fit (Figure 7.3).

However, when the positional tolerance zone is located outside the part and applies to the external projection of the feature (axis) (Figure 7.4), the part fits in any case (Figure 7.5). The external projection has the length of the through-hole in the case of head screws (Figure 7.6) and the length of the outstanding part of the stud or dowel pin, (Figure 7.7). The projected tolerance zone applies only to these lengths, not to the (length of the) axis within the workpiece.

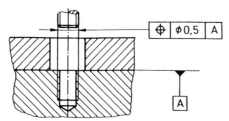

Figure 7.1: Positional tolerance of a threaded hole.

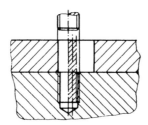

Figure 7.2: Positional tolerance zones of threaded hole and through-hole.

ISO 10 578 defines the projected tolerance zone as follows. The **projected (positional) tolerance zone** applies to the external projection of the feature indicated

Projected Tolerance Zone

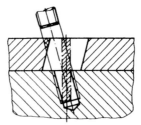

Figure 7.3: Orientation deviations taking advantage of the positional tolerance zones.

on the drawing by the symbol Ⓟ placed in the tolerance frame after the positional tolerance of the toleranced feature. The minimum extent and the location of the projected tolerance zone are shown in the corresponding drawing view by Ⓟ preceding the projected dimension (Figure 7.4).

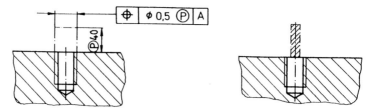

Figure 7.4: Projected tolerance zone.

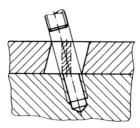

Figure 7.5: Taking advantage of the projected tolerance zone by inclined axes.

The exact definition of the axis to be extended (projected) is not yet standardized (see 3.5). From a practical point of view, the following definitions of the axis may be applicable:

● for plain holes (e.g. interference fits), the axis of the contacting cylinder;

74

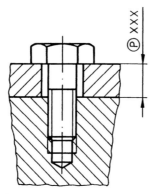

Figure 7.6: Projected tolerance zone for threaded hole and head screw.

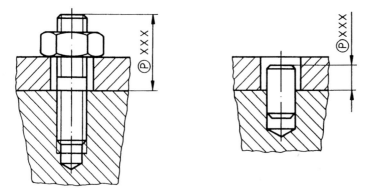

Figure 7.7: Projected tolerance zone for threaded hole and stud or for hole and pin.

● for threaded holes; the axis of a (nearly) geometrical ideal screw of maximum material size (go gauge).

8 Substitute Elements

8.1 GENERAL

Substitute elements (substitute features) are imaginary geometrical ideal features (e.g. straight line, circle, plane, cylinder, sphere, cone and torus). Their location, orientation and (if applicable) size are calculated from the assessed points of the workpiece surface.

In general, substitute elements are assessed by coordinate measuring machines or form measuring machines. According to the planned standard ISO 10 360, the assessment is performed in a right-handed Cartesian (rectangular) coordinate system (Figure 8.1) or a right-handed cylindrical coordinate system (Figure 8.2).

Points in space are defined by their coordinates x_0, y_0, z_0 (location vector P) and stored for further data processing (Figure 8.3). Orientations in space are defined by the components E_x, E_y, E_z of the unit vector (orientation vector N or E). It therefore follows that $|N| = \sqrt{(E_x^2 + E_y^2 + E_z^2)} = 1$ (Figures 8.3 and 8.4). The orientation vector is always directed out of the material.

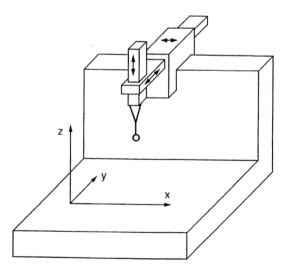

Figure 8.1: Coordinate measuring instrument with Cartesian coordinate system.

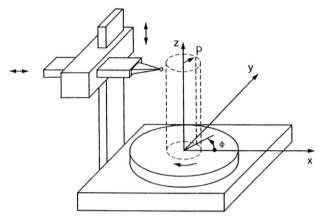

Figure 8.2: Coordinate measuring instrument with cylindrical coordinate system.

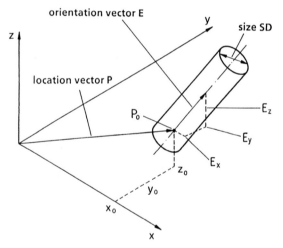

Figure 8.3: Definition of a substitute cylinder (location vector, orientation vector and substitute size).

Table 8.1 shows the vectors needed to define the substitute element. The indicated minimum numbers of points to be assessed apply to geometrical ideal features. As the actual features always exhibit form deviations, many more points are necessary for measuring an actual feature of the workpiece. The workpiece is sensed at a number of points, and, after tracer radius correction, the calculated

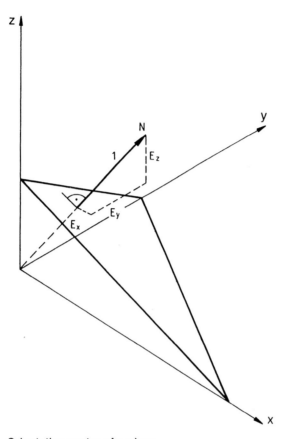

Figure 8.4: Orientation vector of a plane.

coordinates of the actual feature points are stored. From the stored coordinates, the substitute element is calculated (Figure 8.5).

The calculation of the substitute element may be done according to the following procedure:

(a) Gauss (the sum of squares of deviations of the sensed points from the substitute element is minimized);

(b) Chebyshev (the maximum deviation of the sensed points from the substitute element is minimized);

Table 8.1: Substitute elements

Element	Location X_0, Y_0, Z_0	Orientation E_x, E_y, E_z	Size	Remarks	Minimum number of points
Plane	×			Point on plane	3
		×		Normal to plane	
Sphere	×			Centre point	4
			×	Diameter	
Cylinder	×			Point on axis	5
		×		Orientation of axis	
			×	Diameter	
Cone	×			Point on axis	6
		×		Orientation of axis	
			×	Cone angle	
Torus	×			Centre point	7
		×		Orientation of axis	
			× ×	Diameters of pipe and ring	
Straight line	×			Point on straight line	2
		×		Orientation of straight line	
Circle	×			Centre point	3
		×		Normal to plane	
			×	Diameter	

(c) contacting element (only for bounded geometrical elements, maximum inscribed geometrical element for holes or minimum circumscribed geometrical element for shafts).

These methods may yield different results, although they use the same sensed points. Only the Gaussian method is always unique. The planned standard ISO 10 360 will probably require that, unless otherwise specified, the Gaussian method is to be applied.

8.2 VECTORIAL DIMENSIONING AND TOLERANCING

Substitute elements are defined by vectors (location vector, orientation vector and size(s)) in a substitute datum system. Dimensioning and tolerancing with the aid of substitute elements is called vectorial dimensioning and tolerancing, VD&T.

For the use of VD&T the following definitions and drawing indications must

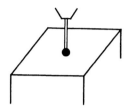

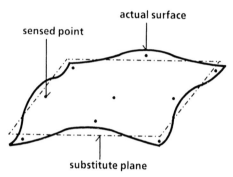

Figure 8.5: Sensed points and substitute element of a plane workpiece surface.

be agreed upon. As there is not yet a standard on this subject (ISO 10360 is not yet sufficient for this purpose), they must be specified in a company standard or by reference to this book.

Actual feature This is the feature (actual surface, actual axis or actual median face) obtained by measurement.

Nominal feature This is the geometrical ideal feature (nominal surface, nominal axis or nominal median face) as defined by the drawing. To identify the feature on the drawing, an enumeration with symbolization according to Figure 8.6 may be used.

$$\phi\!\!-\ 2$$

Figure 8.6: Identification of a feature (No. 2) on the drawing.

Substitute element This is the geometrical ideal element of similar form to the nominal feature (e.g. plane, sphere, cylinder, cone or torus) and derived from the

Substitute Elements

measured data using the Gaussian algorithm (see ISO 10 360).

Substitute axis   This is the axis of a substitute cylinder, cone or torus.

Substitute median plane   This is the median plane between the substitute planes of two opposite parallel plane surfaces.

Substitute intersection point   This is the point of intersection of a substitute axis and a substitute plane. On drawings the substitute intersection point may be symbolized by ⊗ followed by the feature numbers combined by the symbol + (Figure 8.7).

Figure 8.7:   Drawing indication of a substitute intersection point (of substitute elements Nos 2 and 3).

Specified point of a substitute element   This is the point of a substitute element specified to locate the substitute element relative to the substitute datum system. On drawings the specified point may be symbolized by the symbol ⊗ followed by the feature(s) number(s). If the same symbol is applied to more than one feature, the feature numbers are separated by commas (Figure 8.8).

Figure 8.8:   Drawing indication of specified points (of features Nos 5 and 6).

Substitute datum system   This is the coordinate system of the workpiece derived from substitute elements. The location and orientation of the substitute elements are related to this coordinate system. It is defined by the primary, secondary and tertiary substitute datums as follows.

Primary substitute datum (xy *plane*)   This is defined by

- a specified substitute plane, or
- a plane containing the substitute axis of the primary datum feature and, if

applicable, the specified substitute intersection point (apart from the substitute axis, the intersection of another substitute axis with a substitute plane).

Secondary substitute datum (xz plane) This is defined by a plane perpendicular to the primary datum and containing

- the intersection line of the substitute plane of the secondary datum feature with the primary datum, or

- the substitute axis of the primary datum feature (which also thereby becomes a secondary feature; see Figure 8.17), or

- the line between two substitute intersection points specified as secondary datums, this line being not normal to the primary datum.

Tertiary substitute datum (yz plane) This is defined by a plane perpendicular to the primary and to the secondary datums, and containing

- the point of intersection of the substitute plane of the tertiary datum feature with the primary and secondary datum, or

- the substitute intersection point specified as tertiary datum.

To identify a substitute datum for reference purposes, a capital letter preceded by S for substitute is enclosed in a rectangular frame connected to a datum triangle (Figure 8.9).

Figure 8.9: Drawing indication of a substitute datum.

The substitute datum system is indicated by three subsequent rectangular compartments (Figure 8.10). The sequence from left to right indicates the order of primary, secondary and tertiary datum.

Figure 8.10: Drawing indication of a substitute datum system.

If a datum is defined by a combination of an axis and an substitute intersection point or by two substitute intersection points, the datum letters in the datum system frame are combined by a hyphen (Figure 8.11).

SA - SB	

Figure 8.11: Drawing indication of a common substitute datum.

The mathematically positive coordinate system according to ISO 841 applies (Figure 8.12).

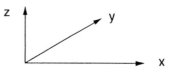

Figure 8.12: Mathematically positive coordinate system.

Substitute datum systems may be specified (e.g. for a pattern of holes) related to the main substitute datum system. There is no accumulation of tolerances related to the two datum systems.

Linear substitute sizes These are the diameters of substitute spheres, cylinders, cones and tori.

The **linear substitute size deviation** is the difference between the actual substitute size and the nominal substitute size.

The **linear substitute size tolerance** is the difference between the upper and lower limits of the linear substitute size deviations.

Angular substitute sizes These are the cone angles of substitute cones.

The **angular substitute size deviation** is the difference between the actual substitute cone angle and the nominal substitute cone angle.

The **angular substitute size tolerance** is the difference between the upper and lower limit of the angular substitute size deviations.

Substitute location This is defined by the **substitute location vector**. It indicates the location of a **specified point** of the substitute element in the substitute datum system. The specified point can be

- the substitute intersection point;

- a point of the substitute element specified to lie in the case of a plane on a straight line defined by the two coordinates and perpendicular to the plane which is defined by the two pertinent coordinate axes (Figure 8.13), or in the case of an axis on a plane defined by one coordinate and parallel to the plane defined by the other two coordinate axes (Figure 8.14);

● the centre of a substitute sphere;

● the centre of a substitute torus.

The **substitute location deviation** is the difference between the actual substitute location vector and the nominal substitute location vector. It can be expressed by its three components in the directions of the axes of the substitute datum system.

The **substitute location tolerance** is the difference between the maximum permissible substitute location deviations expressed in the three directions of the axes of the substitute datum system. In the case of a specified point of a substitute plane two coordinates are theoretically exact, and only one is to be toleranced (Figure 8.13). In the case of a specified point of a substitute axis one coordinate is theoretically exact, and two are to be toleranced (Figure 8.14).

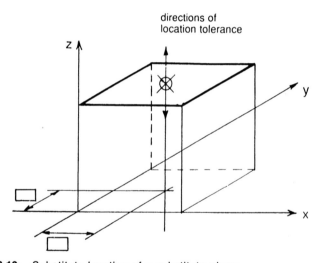

Figure 8.13:   Substitute location of a substitute plane.

Substitute orientation   This is defined by the substitute orientation vector. It indicates the orientation of the substitute plane or the substitute axis within the substitute datum system. The substitute orientation vector is normal to the substitute plane or parallel to the substitute axis (cylinder, cone or torus) and points

● in the case of a plane away from the material;

85

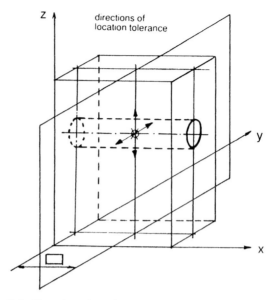

Figure 8.14: Substitute location of a substitute cylinder.

● in the case of a cone in the direction of the greater size.

The **substitute orientation vector** E has magnitude 1 and is defined by its components E_x, E_y, E_z, which are in the directions of the axes of the substitute coordinate system (substitute datum system).

The **substitute orientation deviation** is the difference between the actual substitute orientation vector and the nominal substitute orientation vector. It can be expressed by its three components in the directions of the axes of the substitute datum system.

The **substitute orientation tolerance** is the difference between the maximum permissible substitute orientation deviations expressed in the directions of the axes of the substitute datum system. In order to avoid overspecification only two of the three components of the substitute orientation vectors can be toleranced (the third is already determined by the difference from 1).

Form The nominal form is defined by the drawing or by indication in a table. In the case of a bounded form (sphere, cylinder, cone and torus) and indication in a table it should be stated whether it is an internal form (hole) H or an external form (shaft) S.

The **local form deviation** is the distance of the workpiece feature from the substitute element at the considered location. The **total form deviation** is the range of the local form deviations (see Figure 8.15).

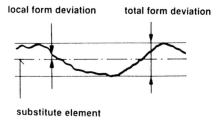

Figure 8.15: Local form deviation and total form deviation.

The **form tolerance** is the value of the permitted total form deviation.

Note that according to ISO 1101, the form tolerance zone has the direction of the minimum zone. With vectorial tolerancing, the form tolerance zone has the direction of the substitute element. There may be small differences in the measurement results between the two methods.

Surface roughness The same applies as with conventional dimensioning and tolerancing. See ISO 4287 P.1 and ISO 4288.

VD&T requires the following specifications:

- substitute datum system (substitute elements and substitute intersection points);

- location and locational tolerance of one point (specified point) of the substitute element (Table 8.1);

- orientation and orientational tolerance of the substitute element (Table 8.1);

- size(s) and size tolerance(s) of the substitute element (Table 8.1);

- cone angle and cone angle tolerance of the substitute cone (Table 8.1);

- form tolerance (superimposed on the substitute element).

These specifications may be listed according to their feature numbers (Figure 8.16).

Substitute Elements

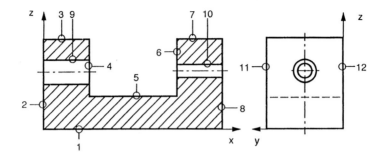

Substitute Element No			1	2	3	4	5	6
Location	Nom.	x_0 y_0 z_0						
	Perm. Dev. ±	x y z						
Orientation	Nom.	E_x E_y E_z						
	Perm. Dev. ± x 10⁻³	e_x e_y e_z						
Size	Nom.							
	Perm. Dev. ±							
Form	Nom.							
	Tol.							

Figure 8.16: Principle of a drawing with vectorial sizes and tolerances.

Figure 8.17 shows an example of VD&T. The workpiece coordinate system is determined by

- the axis of substitute feature No.1, the substitute intersection point of axis of substitute feature No. 2 and the substitute feature No. 3 (SA-SB) to establish the *xy* plane;

- the axis of substitute feature No. 1 (SA), perpendicular to the *xy* plane to establish the *xz* plane;

- the substitute intersection point of the axis of substitute feature No. 2 and substitute feature No. 3 (SB), perpendicular to the *xy* and *xz* planes to establish the *yz* plane.

With the enumeration of specified points the sign "+" (e.g. "2 + 3") indicates that it is the substitute intersection point of two substitute elements, the sign ","

88

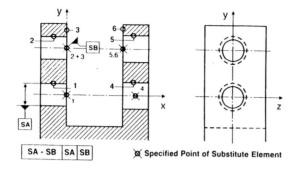

| SA - SB | SA | SB |

⊠ Specified Point of Substitute Element

Primary Datum	1 - 2 + 3	(Substitute El. and Substitute Intersection Point)
Secondary Datum	1	(Substitute Element)
Tertiary Datum	2 + 3	(Substitute Intersection Point)

Substitute Element No			1	2	3	4	5	6
Location	Nom.	x_0	0	0	0	55	45	45
		y_0	0	35	35	0	35	35
		z_0	0	0	0	0	0	0
	Perm. Dev. ±	x	–	–	–	–	–	0,1
		y	–	0,1	–	0,1	0,1	–
		z	–	–	–	0,1	0,1	–
Orientation	Nom.	E_x	–1	–1	+ 1	+ 1	+ 1	–1
		E_y	0	0	0	0	0	0
		E_z	0	0	0	0	0	0
	Perm. Dev. ±	e_x	–	–	–	–	–	–
	x 10-3	e_y	–	0,2	0,2	0,2	0,2	0,2
		e_z	–	0,2	0,2	0,2	0,2	0,2
Size	Nom.		15	15		20	20	
	Perm. Dev. ±		0,05	0,05		0,05	0,05	
Form	Nom.		Zyl. H	Zyl. H	Plane	Zyl. H	Zyl. H	Plane
	Tol.		0,005	0,005	0,01	0,005	0,005	0,001
Waviness	W_t μm		–	–	–	–	–	–
Roughness	R_z μm		6	6	16	6	6	16
Further Requirements			X	X		X	X	

No	Further Requirements
1	MMVC DIA 14.94 rel. to No 4 MMVC DIA 19.94
2	MMVC DIA 14.94 rel. to No 5 MMVC DIA 19.94
4	See No 1
5	See No 2

Figure 8.17: Example of vectorial dimensioning and tolerancing.

(e.g. "5, 6") indicates that there are two different specified points of the same nominal location.

The location tolerances apply to the specified points of the substitute elements, for example for substitute element No. 2 determined by the specified point 2 + 3 in the y direction (because in the z direction it coincides by definition of the workpiece coordinate system with the coordinate $z = 0$) and for substitute element No. 5, determined by the specified point 5 in the y and z directions (in the x direction the point is specified by the nominal value 45 which determines the point of the substitute element axis at which the location deviation should be assessed).

The orientation tolerances apply to the substitute elements themselves (not to specified points). The axis of substitute element No. 1 has no tolerance, because it establishes the x direction of the workpiece coordinate system.

Regarding orientation, the permissible deviations are only specified for two components of the orientation vector, because the permissible deviation of the third component is already determined by the condition that the orientation vector (resultant of the three components) has the value 1.

When the nominal orientation vector is not parallel to any axis of the coordinate system, permissible deviations may be specified for two components only or for all three components. When permissible deviations are specified for all three components, the extreme deviations cannot occur on the same workpiece, because the resultant of the three actual components has the value 1 (by definition).

With form the indication "Cyl H" indicates a cylindrical hole, while "Cyl S" indicates a cylindrical shaft.

In the case of a cylindrical shape it might be useful for process control to distinguish between "conical form deviation" and other (e.g. irregular) form deviations. Then, instead of a cylinder, the form should be specified as a cone of cone angle $0°$ (and the substitute diameter at half of the feature's length). The deviations will then be separated for conicity (in grad) and other form deviations (in mm).

Further requirements may have to be specified, e.g. the envelope requirement Ⓔ according to ISO 8015, the maximum material requirement Ⓜ or the least material requirement Ⓛ according to ISO 2692. In these cases the geometrical ideal boundary of maximum material virtual size or of least material virtual size may be stated using the following symbols:

DIA   diameter;

DIS   distance of two opposite parallel planes;

E   boundary for the envelope requirement (virtual condition) (see 10);

M   boundary for the maximum material requirement (maximum material virtual condition) (see 9);

L   boundary for the least material requirement (least material virtual condition) (see 11).

The theoretical exact dimensions to determine the locations of the virtual conditions (in cases of M or L) are given by, or are to be derived from, the nominal location vectors of the substitute elements. The orientation of the virtual conditions are given by the nominal orientation vectors.

Example 1

Indication: No. 7 E DIA 19.9

Meaning: The actual feature No. 5 should not violate the geometrical ideal boundary (cylinder) of ∅ 19.9.

Example 2

Indication: No. 1 M DIA 14.94 rel. to No. 4 M DIA 19.94

Meaning: The actual features No. 1 and No. 4 should not violate the geometrically ideal boundaries (cylinders, maximum material virtual condition) of ∅ 14.94. The location and orientation of the boundaries are theoretically exact (coaxial) as given by the location and orientation vectors. See Figure 8.17.

Example 3

Indication: No. 8 - No. 9 M DIS 20 rel. to No. 10 - No. 11 M DIS 25
No. 9 see No. 8
No. 10 see No. 8
No. 11 see No. 8

Meaning: The actual features Nos 8, 9 and Nos 10, 11 should not violate the geometrically ideal boundaries (two opposite parallel planes, maximum material virtual condition) of distances 20 and 25 respectively. The location and orientation of the boundaries are theoretically exact (coplanar median planes) and to be derived from the location and orientation vectors.

Further requirements may concern surface roughness, surface treatment conditions etc.

It is also possible to use vectorial tolerancing (of substitute elements) for certain selected features while sizes and tolerances according to ISO 8015, ISO 286, ISO 1947 and ISO 1101 (conventional dimensioning and tolerancing) are used for the other features (Figure 8.18). However, in order to avoid misunderstanding, clear distinctions must exist in the drawing specifications regarding sizes, distances, locations and orientations of substitute elements and their tolerances on the one hand, and sizes, distances and their tolerances and orientational and locational tolerances according to ISO 8015, ISO 286, ISO 1947 and ISO 1101 on the other. Such distinctions in drawing specifications are yet to be standardized. For the time being, they must be specified, e.g. by a company standard or by reference to this book. Figure 8.18 gives an example of an application. The symbols used are similar to ISO 1101.

As form deviations are eliminated by calculation from the substitute elements, this method provides a clear distinction between substitute size, substitute location,

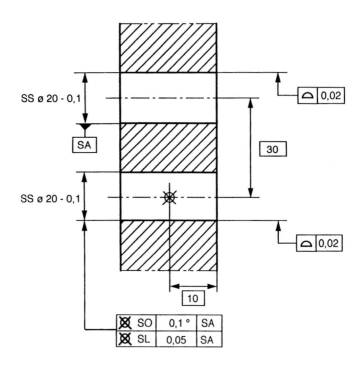

Figure 8.18: Individual vectorial tolerancing of selected features next to (conventional) tolerancing according to ISO 8015, ISO 286 and ISO 1101 in the same drawing. Symbols for substitute elements are similar to ISO 1101.

substitute orientation and form. (Figures 8.19 and 8.20 compare with Figure 3.12.) It further provides the direction of the orientation and location deviation, and therefore gives the appropriate information for directing of the correction of the production equipment (machine tool).

Therefore vectorial dimensioning and tolerancing may be advantageous for specifying some functional requirements, such as for gears, gear boxes, hydro-dynamic bearings and air lubricated bearings, and may be necessary for manufacturing process control (analysing or controlling the manufacturing process).

As deviations in size, orientation and location according to ISO 8015, ISO 1947 and ISO 1101 always include form deviations, they may be less suitable for some of these purposes.

The differences in the measurement results between the two types of sizes and orientation and location deviations from the same workpiece have been found to be up to 100%.

Which method of dimensioning and tolerancing should be used depends on the functional requirements and on the requirements of the manufacturing process control.

8.3 COMPARISON OF SYSTEMS

According to the planned standard ISO 10 360, the sizes, distances and deviations of orientation and location relate to the substitute elements and not to the actual features exhibiting form deviations. Therefore they are designated as substitute sizes, substitute distances and substitute deviations (vectorial system). They are to be distinguished from sizes, distances and deviations according to ISO 8015, ISO 286, ISO 1947 and ISO 1101 (conventional system). In the following the differences are described.

8.3.1 Sizes

Figure 8.21 shows a cylinder with form deviations. The actual local sizes according to ISO 8015 and ISO 268 vary between the minimum and maximum actual local sizes. The actual substitute size in this case is the diameter of the substitute cylinder.

Figure 8.22 shows a cone with form deviations. The actual local cone angles according to ISO 8015 and ISO 1947 vary between the maximum and minimum actual local cone angles. The actual substitute cone angle is the cone angle of the substitute cone (which deviates from the nominal cone angle).

Figure 8.23 shows a bar with holes. The bar has deviations from straightness. The actual local distances between the edge and the hole centres according to ISO 8015 are approximately the same at all locations. The distances of the hole centres from the substitute plane, however, differ throughout.

Substitute Elements

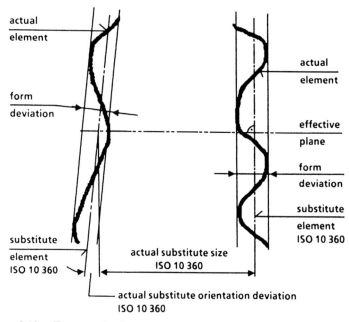

Figure 8.19: Two nominally parallel surfaces: determination of substitute size, form, substitute orientation and substitute location separately from each other (by using substitute elements).

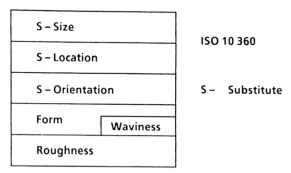

Figure 8.20: Types of deviations and tolerances of substitute elements (cf. Figure 3.12).

Depending on the data processing program incorporated, the coordinate measuring machine can also assess actual local sizes (linear sizes and angular sizes) according to ISO 8015.

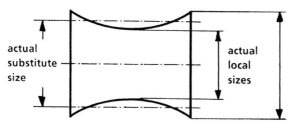

Figure 8.21: Cylinder with form deviations, actual local sizes and actual substitute size.

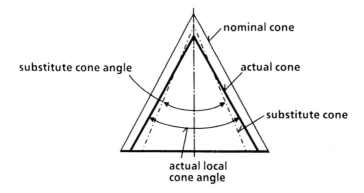

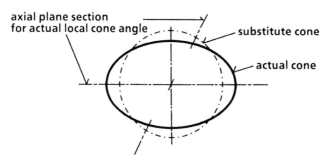

Figure 8.22: Cone with form deviations, actual local cone angles and actual substitute cone angle.

However, in this case, difficulties arise from the fact that an unambiguous standardized definition of the actual local size does not yet exist. Therefore at present coordinate measuring machines use different definitions of actual sizes.

95

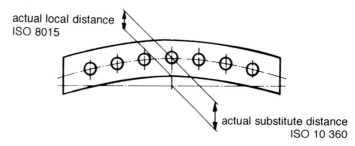

actual local distance
ISO 8015

actual substitute distance
ISO 10 360

Figure 8.23: Bar with holes and deviations from straightness, actual local distances and distances from the substitute plane.

The distance between two parallel planes, for example, is assessed as (Figure 8.24)

(a) distance perpendicular to a datum substitute plane to be chosen;

(b) distance between the corners of the substitute planes;

(c) distance between the centre of gravity S of one substitute plane and the other substitute plane;

(d) distance between two parallel substitute planes.

The procedures for assessing actual local sizes that are planned to be standard-ized in an ISO Standard are as follows.

(a) Distance between two parallel faces (e.g. wall thicknesses) The actual local size is the distance between two opposite points, the connection of the two points being perpendicular to the substitute median plane. The substitute median plane is the median plane between the substitute planes of the two opposite surfaces (Fig. 8.25).

(b) Diameter The actual local diameter (size) is the distance of two opposite points estimated in a section perpendicular to the substitute cylinder axis or to the substitute cone axis, the line connecting the two points meeting the centre of the Gaussian regression (least-squares) circle in this plane (Figure 8.26). The substitute cone is the best-fit cone; the cone angle may deviate from the nominal cone angle.

It must be noted, however, that these definitions have not yet been standardized. Opinions still differ as to which substitute circle should be used. The Gaussian regression circle has the advantages of needing the least number of traced points and of always being unique. The Chebyshev substitute circle has the advantage of being standardized in ISO 1101 for the assessment of roundness, but the disadvantages of needing a much larger number of traced points and not always

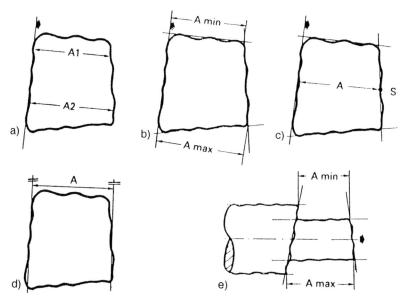

Figure 8.24: Coordinate measuring technique: former procedures to assess the distance between two parallel surfaces (according to Wirtz).

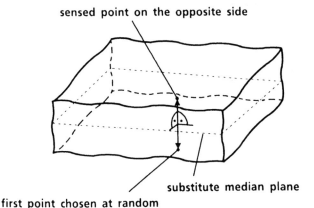

sensed point on the opposite side

substitute median plane

first point chosen at random

Figure 8.25: Measurement of thickness; actual local size.

being unique. The contacting substitute circle (maximum inscribed or minimum circumscribed) has the advantage of being in conformance with ISO 5459 for the definition of datums, but has the disadvantage of not always being unique. The author recommends the use of the least-squares method (see 3.5).

97

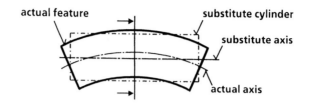

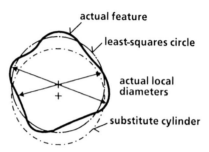

Figure 8.26: Measurement of actual local diameter.

8.3.2 Form deviations

When the Chebyshev criterion (see 3.5 and 8.1) is used for obtaining a substitute element and the form deviation is estimated in relation to this substitute element, the approach is the same as according to ISO 1101.

If other criteria are used, the value of the form deviation obtained may be larger.

Old coordinate measuring machines sometimes measure the form deviation as the maximum distance of the traced points from the substitute element (Figure 8.27). This is then only half the value of the form deviation according to ISO 1101.

The distance of the traced point from the substitute element is the actual local form deviation (Figure 8.28). The form deviation according to ISO 1101 is the range of the actual local form deviations.

8.3.3 Deviations of orientation and location

With coordinate measuring machines, deviations of orientation are normally obtained as an inclination of the substitute element (Figure 8.29). In this procedure the form deviation is eliminated. The sign of the orientation deviation indicates

(a)

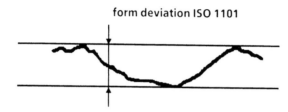

(b)

Figure 8.27:   Form deviation: (a) former coordinate measurement; (b) according to ISO 1101.

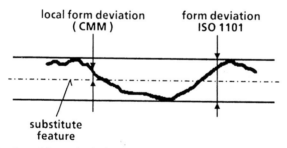

Figure 8.28:   Local form deviation.

the direction of the orientation deviation. These are the main differences from the deviation of orientation according to ISO 1101.

Figure 8.30 shows a workpiece with deviations of coaxiality. The deviation of coaxiality according to ISO 1101 also includes the form deviation of the axis of the cylinder to be measured, and is therefore greater than the coaxiality deviation of the substitute element.

Figure 8.31 shows a workpiece with a coaxiality deviation according to ISO 1101 and practically no coaxiality deviation of the substitute element.

Substitute Elements

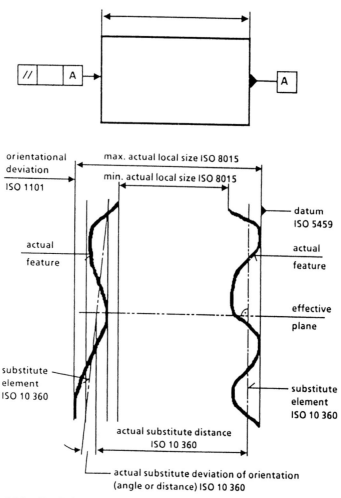

Figure 8.29: Deviation of orientation (parallelism) according to ISO 1101, and orientational deviation of the substitute element.

Figure 8.32 shows a workpiece with a positional deviation. The positional deviation according to ISO 1101 is greater than the positional deviation of the substitute elements.

The positional deviation according to ISO 1101 includes the form deviation of the toleranced feature, and is related to the datum that contacts the datum feature.

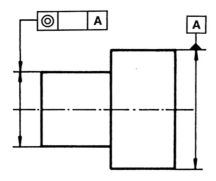

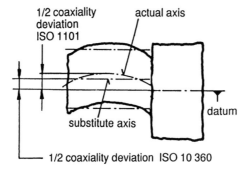

Figure 8.30: Coaxiality deviation according to ISO 1101, and coaxiality deviation of the substitute element.

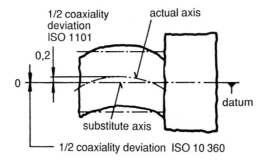

Figure 8.31: Workpiece with a coaxiality deviation according to ISO 1101 but with practically no coaxiality deviation of the substitute element.

101

Substitute Elements

The positional deviations of the substitute elements differ from those according to ISO 1101 because the form deviations of both features (the toleranced feature and the datum feature) are eliminated when the positional deviation of the substitute element is assessed and the substitute datum element intersects the actual datum feature.

When appropriate data processing programs are available, coordinate measuring machines can also calculate deviations of orientation or location according to ISO 1101.

However, it is therefore necessary to specify the definition of the actual axis of a cylindrical or conical feature. According to ISO 1101, it is assumed that the actual axis follows the form of the feature and therefore deviates from a straight line (see 3.5).

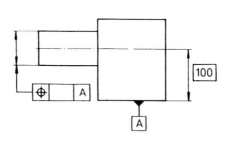

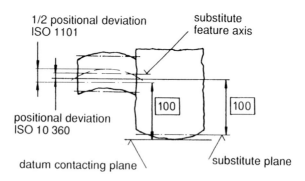

Figure 8.32: Positional deviation according to ISO 1101, and positional deviation of the substitute element.

8.3.4 Datum systems

According to ISO 10 360, the standard is in preparation, the datum systems for workpieces for the determination of location and orientation of the substitute elements (substitute datum system) may be defined as follows (see also 8.2: substitute datum system).

Primary datum This is the substitute plane of the primary datum feature (xy plane, if not otherwise specified).

Secondary datum This is the plane normal to the primary datum containing the line of intersection of the latter and the substitute plane of the secondary datum feature (xz plane, if not otherwise specified).

Tertiary datum This is the plane normal to the primary datum and to the secondary datum containing the point of intersection of all three substitute planes (yz plane, if not otherwise specified).

This datum system for substitute elements is based on substitute planes and their intersections. Workpiece datum systems according to ISO 1101 and ISO 5459, however, are based on contacting elements contacting the highest points of the workpiece surfaces and directed according to the minimum-rock requirement.

In order to obtain comparable results using coordinate measuring machines and measuring deviations according to ISO 1101, the substitute datum system must be transformed.

Two options are available for this transformation. One involves moving the three coordinate system planes normally to themselves until they contact the highest sensed points of the surfaces (Figure 8.33). The other is to move them normally to themselves by half of the form deviation plus half of the orientational deviation.

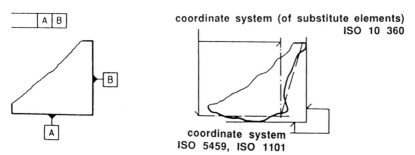

Figure 8.33: Datum system derived from substitute elements according to ISO 10360 and transformed to the highest sensed points.

In the first case a measurement strategy is needed that confines sensing to the critical areas containing the highest points. A guideline for this strategy does not yet exist.

In the second case the amount of transformation calculated may be too large when the surface is convex.

However, in general, the resulting error may be negligible when a sufficient number of points are sensed. It is important that the coordinate measuring machine can achieve this transformation in order to avoid the described systematic error between datum systems according to ISO 10 360 and according to ISO 5459. Otherwise deviations of location according to ISO 1101 obtained with coordinate measuring machines will be wrong.

8.4 CONVERSION BETWEEN SYSTEMS, COMBINATION OF SYSTEMS

Both systems (the conventional one and the vectorial one) have advantages and disadvantages. For example, the vectorial system has advantages for manufacturing process control, while the conventional system has advantages for specifying the functional requirements in the cases of clearance fits.

It is likely that algorithms will be developed that for particular features convert conventional tolerances (e.g. according to ISO 1101) into vectorial tolerances, and perherbs also vice versa.

So in one layer of the workpiece data the functional requirements (for assembly) may be indicated by the profile of surface tolerances related to a workpiece datum system. In another layer of the workpiece data the vectorial tolerances (for manufacturing control) related to the same workpiece datum system may be indicated. For both indications the same nominal data (location and orientation vectors) apply. See also 19.2.

9 Maximum Material Requirement

9.1 DEFINITIONS

Actual local size   This is any individual distance at any cross-section of a feature, i.e. any size measured between any two opposite points (two-point measurement) (ISO 286, ISO 2692) (Figure 9.1).

Each feature of an individual workpiece theoretically has an infinite number of actual local sizes.

Exact definitions of actual local sizes are not yet standardized. The problems are the definitions of "opposite" and of the directions of the cross-sections (see 8.3.1).

Maximum material condition (MMC):   This is the state of the considered feature in which the feature is everywhere at that limit of size where the material of the feature is at its maximum, e.g. minimum hole diameter and maximum shaft diameter (ISO 2692). The actual axis of the feature need not to be straight.

Mating size for an external feature: This is the dimension of the smallest perfect feature (e.g. a cylinder or two parallel opposite planes) that can be circumscribed about the feature so that it just contacts the surface at the highest points (Figure 9.1) (ISO 2692).

Mating size for an internal feature   This is the dimension of the largest perfect feature (e.g. a cylinder or two parallel opposite planes) that can be inscribed within the feature so that it just contacts the surface at the highest points (ISO 2692).

Maximum material size (MMS):   This is the dimension defining the maximum material condition of a feature (Figure 9.1) (ISO 2692), i.e. the limit of size where the material is at the maximum, e.g. the maximum limit of the size of a shaft or the minimum limit of the size of a hole.

Maximum material virtual size (MMVS):   This is the size generated by the collective effect (concerning mating) of the maximum material size (MMS) and

105

the geometrical tolerance followed by the symbol Ⓜ, i.e.

for shafts MMVS = MMS + geometrical tolerance,
for holes MMVS = MMS − geometrical tolerance.

The MMVS represents the design dimension of the functional gauge (ISO 2692).

Maximum material virtual condition (MMVC): This comprises the features limiting boundary of perfect (geometrical ideal) form and of maximum material virtual size (MMVS) (Figure 9.1). When more than one feature or one or more datum features are applied to the geometrical tolerance, the MMVS are in the theoretical exact locations and orientations relative to each other (ISO 2692).

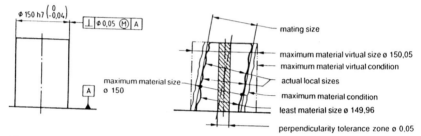

Figure 9.1: Sizes and conditions of a feature with a related geometrical tolerance.

9.2 DESCRIPTION OF MAXIMUM MATERIAL REQUIREMENT

Maximum material requirement (MMR)   This is the requirement, indicated on drawings by the symbol Ⓜ placed after the geometrical tolerance of the toleranced feature or after the datum letter in the tolerance frame, that specifies the following:

● when applied to the toleranced feature, the maximum material virtual condition (MMVS) of the toleranced feature should not be violated (see 9.1);

● when applied to the datum, the related maximum material virtual condition (MMVS) of the datum feature should not be violated.

The size of the related maximum material virtual condition of the datum feature is

● the maximum material size when the datum has no geometrical tolerance followed by the symbol Ⓜ (Figures 9.2 and 9.3.) (this has the same effect as the indication Ⓔ behind the size tolerance of the datum feature);

- maximum material size + (for shafts) or − (for holes) the geometrical tolerance followed by the symbol Ⓜ and applied to the datum, when this tolerance frame is connected with the datum triangle (Figure 9.4).

In the case of Figure 9.5 a symmetry tolerance applies to the datum B. However, since the tolerance frame is not connected to the datum triangle of B, this tolerance does not contribute to the MMVS of the datum B related to the positional tolerance of the four holes.

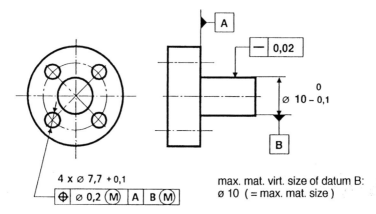

Figure 9.2: Maximum material virtual size of datum B; form tolerance (straightness) to be disregarded.

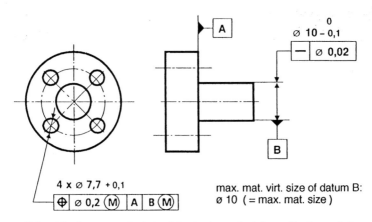

Figure 9.3: Maximum material virtual size of datum B; form tolerance (straightness) to be disregarded.

Maximum Material Requirement

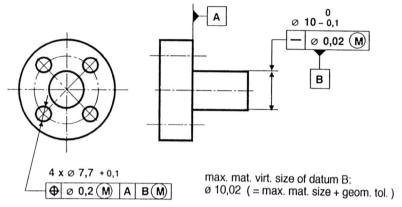

Figure 9.4: Maximum material virtual size of datum B; form tolerance (straightness) to be regarded.

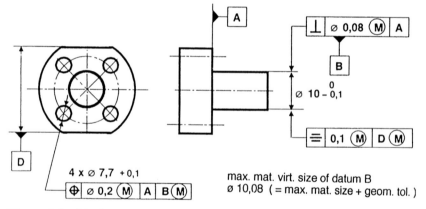

Figure 9.5: Maximum material virtual size of datum B; perpendicularity tolerance to be considered but symmetry tolerance to be disregarded.

For the MMVS of a datum only those geometrical tolerances come into consideration that are related to the datums of the considered tolerance frame (Figure 21.9). Geometrical tolerances of the datum feature followed by Ⓜ that are related to other features do not come into considerations for the MMVS (Figure 21.11). In order to make it more obvious which tolerances followed by Ⓜ contribute to the MMVS of the datum, the planned new version of ISO 2692

108

will specify that those that contribute must have the tolerance frame connected to the datum triangle (Figures 9.4, 9.5 and 9.15).

In other words, the only geometrical tolerance contributing to the MMVS of the datum is that whose tolerance frame is connected with the datum triangle. At present this rule is planned for ISO 2692 but not yet finally accepted. In any case it is recommended that drawing indications be chosen according to this rule.

The maximum material requirement can be explained as a (functional) gauging requirement. The maximum material virtual condition at the toleranced feature and at the datum(s) describe the theoretical form, the theoretical sizes and the theoretical locations and orientations of the gauge surfaces. The workpiece must fit into this gauge. The gauge also represents the most unfavourable counterpart. When the workpiece fits into the gauge, it also fits into all counterparts. The maximum material virtual condition is the (imaginary and geometrical ideal) boundary that must not be violated.

The description given above applies or is planned acording to ISO 2692. ANSI 14.5M deviates from ISO 2692 to a certain extent (see 21.1).

9.3   APPLICATION OF MAXIMUM MATERIAL REQUIREMENT

9.3.1   General

The maximum material requirement can be applied only to those features with an axis or a median plane (cylindrical features or features composed of two opposite parallel planes) (Table 9.1). It cannot be applied to a plane surface or a line on a surface.

The maximum material requirement can be applied when there is a functional relationship between size and form or size and orientation or size and location, i.e. when the geometrical deviation may be larger if the size deviation is smaller. This applies normally to clearance fits. For transition fits and interference fits and for kinematic linkages (e.g. distances of axes of gears) the maximum material requirement is normally not appropriate. This because enlarging the geometrical tolerance (when the size tolerance is not fully used) is detrimental to the function of the part.

9.3.2   Maximum material requirement for the toleranced feature

The maximum material requirement for the toleranced feature allows an increase in the geometrical tolerance when the feature deviates from its maximum material condition (in the direction of the least material condition), provided that the maximum material virtual condition (gauge boundary) is not violated (Figures 9.6–9.11). That is, the maximum material requirement specifies that the indicated geometrical tolerance applies when the feature is in its maximum material condition (largest shaft, smallest hole). When the feature deviates from the

109

Table 9.1: Possible applications of Ⓜ and Ⓛ

	Tolerance	Symbol	Toleranced feature					Datum feature				
			Section line	Edge	Axis	Median face	Surface	Section line	Edge	Axis	Median face	Surface
Unrelated tolerance	Line profile	⌒	X	X	X							
	Straightness	—	X	X	X							
	Roundness	○	X	X								
	Surface profile	⌓				X	X					
	Flatness	□				X	X					
	Cylindricity	⌭					X					
Related tolerance	Line profile with datum(s)	⌒	X	X				X	X	X	X	X
	Surface profile with datum(s)	⌓				X	X	X	X	X	X	X
	Inclination	∠	X*	X	X	X	X	X*	X	X	X	X
	Parallelism	//	X*	X	X	X	X	X*	X	X	X	X
	Perpendicularity	⊥	X*	X	X	X	X	X*	X	X	X	X
	Position	⊕		X	X	X	X		X	X	X	X
	Coaxiality	◎			X					X		
	Symmetry	≡			X	X				X	X	
	Circular run-out	⟋	X	X						X		
	Total run-out	⟭		X			X			X		

maximum material condition (thinner shaft, larger hole), the geometrical deviation may be larger without endangering the mating capability.

The maximum material virtual condition represents the theoretical functional gauge at the toleranced feature. The maximum material virtual size represents the theoretical size of the functional gauge.

Figure 9.6 shows a bolt that is to fit into a hole. According to the definition of the maximum material requirement, in both cases (bolt and hole) the maximum material virtual size (gauge size) is ⌀ 20 (Figure 9.7). When the bolt is everywhere at its maximum material size ⌀ 19.99, the straightness deviation of its axis may be 0.01, as indicated in the tolerance frame (Figure 9.8). When the bolt is thinner, the straightness deviation may be larger. In the extreme case when the bolt is everywhere at its least material size, the straightness deviation may be 0.01 + 0.01 = 0.02 (Figure 9.8). In any case the bolt fits into the gauge and therefore into the most unfavourable counterpart (hole). For the counterpart (hole) the corresponding applies, i.e. the gauge must fit into the hole. Therefore both parts fit.

Figure 9.9 shows two parts that are to fit together. From GD&T and according to the definition of the maximum material requirement, the maximum material virtual condition (gauge boundary) given in Figure 9.10 derived.

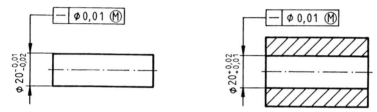

Figure 9.6: Maximum material requirement applied to the straightness tolerance of the axis.

Figure 9.7: Gauge boundary for GD&T according to Figure 9.6.

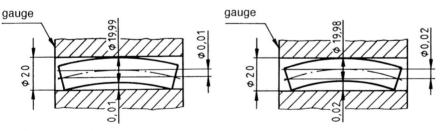

Figure 9.8: Permissible extreme straightness deviations of the bolt according to Figure 9.6.

Maximum Material Requirement

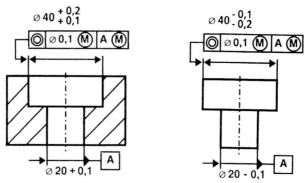

Figure 9.9: Maximum material requirement for part and counterpart.

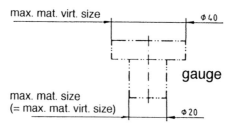

Figure 9.10: Gauge for GD&T according to Figure 9.9.

When the shafts have everywhere the maximum material sizes ⌀ 39.9 and ⌀ 20 (and geometrical ideal form), the coaxiality deviation may be 0.05, which corresponds to a coaxiality tolerance of ⌀ 0.1, as indicated in the tolerance frame in Figure 9.9 (see 18.6). When the bolt head is thinner, the coaxiality deviation may be larger. In the extreme case when the bolt head has everywhere the least material size ⌀ 39.8 (and geometrical ideal form), the coaxiality deviation may be 0.1, which corresponds to a coaxiality tolerance of ⌀ 0.2 (i.e. $0.1 + 0.1 = 0.2$). In any case the bolt fits into the gauge and therefore also into the most unfavourable counterpart. For the counterpart (the two holes) the corresponding applies, i.e. the gauge must fit into the holes.

Figure 9.11 shows a pattern of holes to fit with bolts. The geometrical ideal positions of the hole axes are determined by the theoretical exact dimensions (in rectangular frames). The positional tolerance ⌀ 0.2 of the holes is derived from the maximum material sizes, i.e. the maximum bolt ⌀ 3 and the minimum hole ⌀ 3.2 as described in 6.7.

When the holes have maximum material size ⌀ 3.2 and their actual axes are at maximum permissible separation (i.e. at the border of the tolerance zone ⌀ 0.2

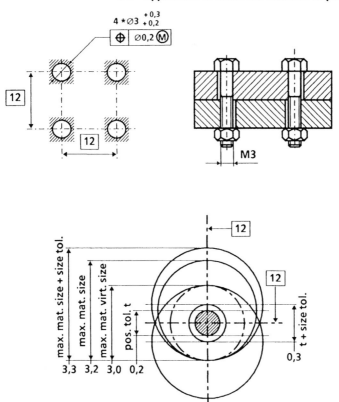

Figure 9.11: Maximum material requirement Ⓜ at the toleranced feature.

on opposite sides), the holes still provide free space for the thickest bolt. When the holes become larger than ∅ 3.2 (e.g. ∅ 3.3), the positional tolerance ∅ 0.2 can be increased by the amount the hole is enlarged (0.1) to become ∅ 0.3. This allows the actual axes of the holes to be 0.3 apart from each other (or 0.15 apart from the theoretically exact positions). In any case a gauge of maximum material virtual size and theoretical exact position fits with the holes, i.e. the hole surfaces must not violate the maximum material virtual condition (gauge boundary). This is the real criterion of the maximum material requirement.

9.3.3 Maximum material requirement for the datum

The maximum material requirement for the datum permits floating of the datum axis or datum median plane relative to the toleranced features pattern when the datum feature deviates from its maximum material condition (in the direction of the least material condition). A prerequisite is that the datum feature does not violate its maximum material virtual condition, which is geometrically ideally positioned in relation to the geometrical ideal position of the toleranced features. Within this boundary, the datum feature may take, if possible, the position where the requirements at the toleranced features are fulfilled.

The amount of float (width or diameter of the floating zone) is equal to the difference between the maximum material virtual size and the mating size of the datum feature (Figure 9.13). This applies for primary datums as in Figures 9.12 and 9.13. For secondary or tertiary datums, as in Figures 21.8 and 21.9, the amount (size of the floating zone) is smaller because of the perpendicularity deviations that must be taken into account.

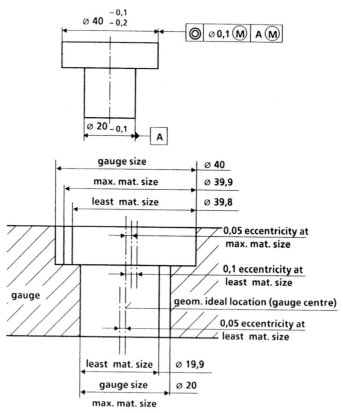

Figure 9.12: Maximum material requirement and maximum possible coaxiality deviation.

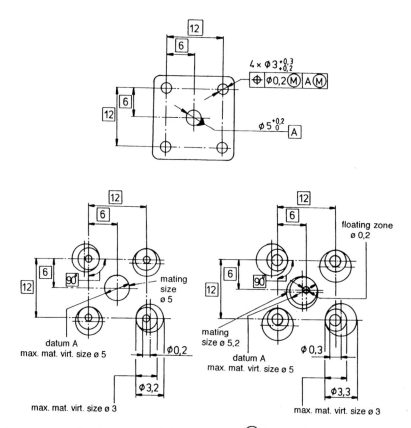

Figure 9.13: Maximum material requirement Ⓜ for the datum.

The deviation of the datum feature from its maximum material virtual size does not increase the tolerance of the toleranced features relative to each other. It only permits a displacement of the pattern of tolerance zones (maximum material virtual conditions of the toleranced features) relative to the actual axis or actual median face of the datum feature (Figure 9.13). However, when only two features are related, as in Figure 9.12, the floating (displacement) has the effect of enlarging the tolerance of the toleranced feature.

115

The functional gauge embodies the maximum material virtual condition of the datum. The maximum material virtual size represents the theoretical size of the gauge. (ISO 2692 - 1988 is still limited to primary datums, which have no form tolerance applied to the axis or median face. In this case the maximum material virtual condition is identical with the maximum material condition (see 9.2).)

Figure 9.12 shows for the example of Figure 9.9 the maximum possible deviation 0.05 of the datum axis relative to the gauge axis (which represents the geometrical ideal position). The maximum possible deviation of the toleranced features (head) axis relative to the gauge axis is 0.1. The maximum possible distance between both bolt axes is 0.15, corresponding to a coaxiality tolerance of $\emptyset$ 0.3 (see 18.6).

Figure 9.13 shows a pattern of holes similar to Figure 9.11 but with a centre hole as datum. The surface of the datum hole must not violate the maximum material virtual condition (gauge boundary) of $\emptyset$ 5.

When the datum hole is everywhere at its maximum material size (and of geometrical ideal form), the positions of the positional tolerance zones and of the maximum material virtual conditions of the four holes relative to the datum hole are determined by the theoretical exact dimension 6.

When the datum hole is larger, it can float within the boundary of the maximum material virtual condition that is represented by the gauge. The diameter of the floating zone within which the actual datum axis can float is equal to the difference between the maximum material virtual size and the mating size of the datum feature. When the mating size of the datum feature is equal to the least material size $\emptyset$ 5.2, the diameter of the floating zone is $\emptyset$ 0.2.

The theoretical exact dimension 6 determines the position of the maximum material virtual conditions (theoretical gauge dimensions). The distances between the actual datum axis and the maximum material virtual conditions of the four holes (gauge axes at the four toleranced holes) may vary, corresponding to the floating zone. This does not alter the magnitudes of the maximum material virtual conditions (gauge dimensions) of the four holes or the distances between them.

All of these requirements can be summarized by the following:

The maximum material virtual conditions of the toleranced features and of the datum feature in their theoretically exact location relative to each other determine the boundary of the functional gauge within which the part must fit.

In Figure 9.14 the additional requirement of perpendicularity relative to the datum surface B is indicated. Accordingly, the maximum material virtual conditions of the toleranced features (four holes) and of the datum A feature (centre hole) are perpendicular to the datum B (plane surface). Here the diameter of the floating zone is not equal to the difference between the maximum material virtual size and the mating size, as in Figure 9.13, but rather to the difference between the maximum material virtual size and the size of the contacting element perpendicular to the datum B (size of the largest inscribed cylinder perpendicular to the datum B). See also Figure 20.36.

This means, in terms of gauging, that the gauge face must be adjusted with the workpiece datum B surface, and then the gauge mandrels must go through the holes (see 20.8.9).

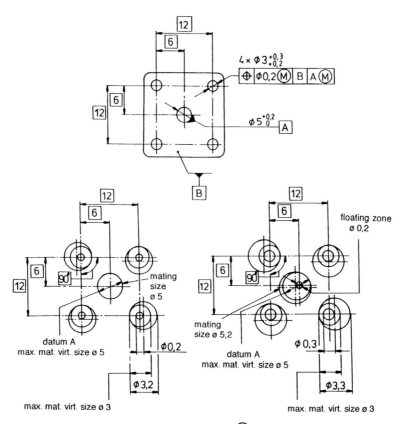

Figure 9.14: Maximum material requirement Ⓜ for the datum, and additional requirement of perpendicularity to the plane surface datum B.

117

Maximum Material Requirement

In Figure 9.15 the maximum material requirement applies to the datum feature controlling the position of the four holes. In addition, for the datum feature itself there applies a straightness tolerance for the hole axis with the maximum material requirement. The maximum material virtual condition of the feature alone (as an isolated feature) is derived from the maximum material size (∅ 5.1) and the straightness tolerance (∅ 0.1). Therefore the maximum material virtual size is ∅ 5. The same maximum material virtual condition applies to the feature as the datum A for the positional tolerance of the four holes. This is indicated by the datum triangle at the tolerance frame of the straightness tolerance of the datum feature.

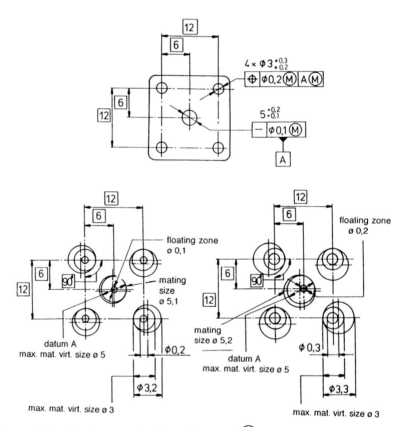

Figure 9.15: Maximum material requirements Ⓜ for the datum and straightness tolerance to be taken into account for the datum.

9.3.4 Maximum material requirement 0 Ⓜ

When the functional tolerance is not distributed on size and position, but is provided for both for random distribution, this is to be indicated on the drawing by 0 Ⓜ (Figure 9.16). Here also, the maximum material virtual condition must not be violated (in Figure 9.16 the cylinder of maximum material virtual size = maximum material size = ∅ 3). See also 9.3.5 and 9.3.6.

Sometimes on drawings the indication according to the right-hand part of Fig. 9.17 appears. This has the same meaning as the indication using the symbol Ⓔ according to ISO 8015 and shown on the left of Figure 9.17. In both cases the envelope requirement applies, i.e. the boundary of the geometrical ideal form and the maximum material size must not be violated. The symbol Ⓔ has been standardized because the drawing indication is simpler than with 0 Ⓜ and in

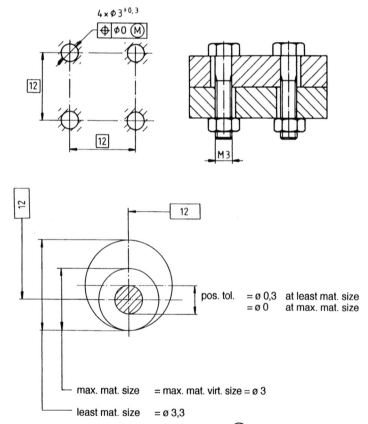

$4 \times \phi 3^{+0.3}$

pos. tol. = ∅ 0,3 at least mat. size
= ∅ 0 at max. mat. size

max. mat. size = max. mat. virt. size = ∅ 3

least mat. size = ∅ 3,3

Figure 9.16: Maximum material requirement 0 Ⓜ.

119

order to avoid difficulties in interpretation according to the chosen geometrical tolerance symbol (see Figure 10.1).

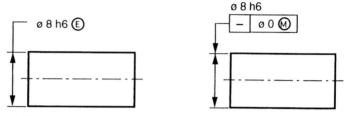

Figure 9.17: Envelope requirement.

9.3.5 Comparison of 0.1 Ⓜ and 0 Ⓜ

Figure 9.18 shows a plug and socket that should fit. The same maximum material virtual conditions (gauging boundaries) apply to both, in cases (a) as well as in

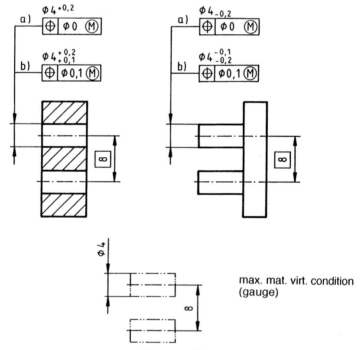

Figure 9.18: Comparison of 0.1 Ⓜ and 0 Ⓜ with the same maximum material virtual condition (gauging boundary).

cases (b). The difference between cases (a) and (b) is the size tolerance and therefore the possible clearence between bolt and hole. In case (a) the clearance may be 0 (hole $\varnothing$ 4 and bolt $\varnothing$ 4). In case (b) the clearance is at least 0.2 but may be utilized by the straightness deviations of the axes (bent hole and bent bolt) (Figure 9.19). In case (b) the manufacturer obtains a recommendation for the distribution of the total provided tolerance (0.2) on size (0.1) and on distance (0.1) (see also 9.3.6).

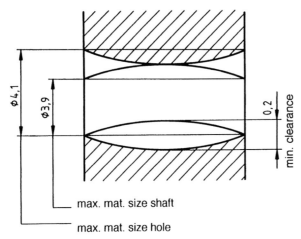

max. mat. size shaft

max. mat. size hole

Figure 9.19: Least clearence (0.2) according to the tolerancing of Figure 9.18 (b) with bent bolt and bent hole.

9.3.6 Reciprocity requirement associated with maximum material requirement

Figure 9.20 shows three fits (a), (b) and (c). In all cases the same maximum material virtual conditions (gauging boundaries) apply for part and counterpart. The differences between cases (a) and (b) are as described in 9.3.4 and 9.3.5. In case (b) the coaxiality tolerance will be enlarged by the size tolerance, not utilized but not vice versa. The size tolerance cannot be enlarged by the non-utilized coaxiality tolerance, although the function (clearance fit) would allow this. To allow this on the drawing, the reciprocity requirement with the symbol Ⓡ after the symbol Ⓜ after the geometrical tolerance should be applied.

Reciprocity requirement associated with the maximum material requirement RR: This requirement, indicated on drawings by the symbol Ⓡ placed after the symbol Ⓜ after the geometrical tolerance in the tolerance frame, specifies that the maximum material virtual condition (MMVC) of the toleranced feature should advantage of the total tolerance (sum of tolerances).

121

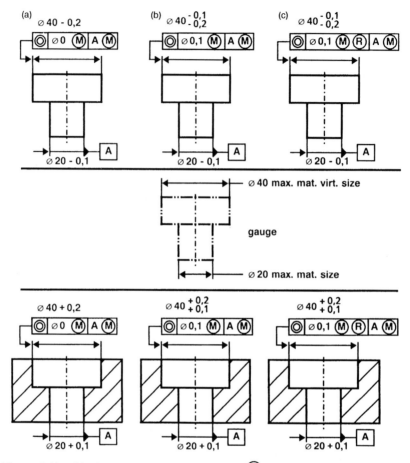

Figure 9.20: Maximum material requirement Ⓜ after geometrical tolerance and after datum letter) and reciprocity requirement Ⓡ after Ⓜ after the geometrical tolerance).

The reciprocity requirement associated with the maximum material requirement has the same effect as 0 Ⓜ, i.e. the total tolerance may be utilized for deviations of size, form, orientation or location in an arbitrary way. However, in contrast to 0 Ⓜ, the drawing indication with the reciprocity requirement Ⓡ gives a recommendation to the manufacturer for the distribution of the total tolerance on size and geometrical characteristics. Thus the reciprocity requirement provides communication between production planning and workshop.

See also 20.11 and Figures 20.48, 20.50, 20.62, 20.64, 20.66 and 20.77.

It was discussed whether the reciprocity requirement should be symbolized with Ⓡ after the size tolerance (because it modifies the latter) or with Ⓜ or Ⓛ after the size tolerance (in order to avoid the introduction of a new symbol) or by Ⓡ after the symbol Ⓜ or Ⓛ after the geometrical tolerance in the tolerance frame. The Draft International Standard ISO DIS 2692 provides an Annex with the latter.

For the reciprocity requirement associated with the least material requirement see 11.4.

9.4 EDUCATION

The fact that the application of the maximum material requirement appears rather complicated sometimes leads to the false opinion that application is not practicable and therefore to be ignored.

However, in many cases the precise functional requirements can only be indicated with the aid of the maximum material requirement. Only then do the largest possible tolerances appear. Therefore this is often unavoidable when economic production is to be achieved.

A prerequisite for application of the maximum material requirement is appropriate education and appropriate planning of the manufacturing and inspection. See also 15.2 and 18.7.12.

For education, in most cases it should be sufficient to explain the maximum material requirement by the following gauging rule:

Where Ⓜ occurs, gauging is required. At the toleranced feature the gauge size is to be calculated in the following way from the maximum material size and the geometrical tolerance, which is followed by the symbol Ⓜ:

for shafts: maximum material size + geometrical tolerance;

for holes: maximum material size − geometrical tolerance.

At the datum feature the gauge size is to be calculated in the same way when at the datum feature a geometrical tolerance followed by the symbol Ⓜ is indicated and the datum triangle is connected directly to this tolerance frame (Figures 9.4 and 9.15).

When there is no geometrical tolerance followed by the symbol Ⓜ indicated at the datum, the gauge size is equal to the maximum material size (Figures 9.2, 9.3 and 9.14).

Designers should know that at the datum (of the original geometrical tolerance) only geometrical tolerances (tolerance frames) should be connected with the datum triangle that have no relationship to other features (straightness of an axis, Figure 9.4) or that are related to datums occuring in the datum system of the original tolerance (Figure 9.5).

According to ANSI Y 14.5 M, there are different rules. Connection of the datum triangle to the considered tolerance frame is not mandatory. The standard requires an analysis of tolerance controls applied to a datum feature in determing the size of the gauge. Further, the standard specifies rules when at the datum feature geometrical tolerances are indicated that are not followed by the symbol Ⓜ. If necessary, see 21.1.2.

10 Envelope Requirement

10.1 DEFINITION

The envelope requirement according to ISO 8015 specifies that the surface of a single feature of size (cylindrical surface or a feature established by two parallel opposite plane surfaces) should not violate the imaginary envelope of perfect (geometrical ideal) form at maximum material size

The envelope requirement may be specified either

- by the symbol Ⓔ placed after the linear (size) tolerance, when applicable to an selected individual feature;

- by indication in the drawing title box "ISO 2768 ...E", when applicable to all features of size in addition to the general geometrical tolerances according to ISO 2768;

- by a national standard (e.g. ANSI Y14.5M or DIN 7167), when applicable to all features of size.

The envelope requirement cannot be applied to features for which a straightness or flatness tolerance is specified that is larger than the size tolerance. According to ANSI Y14.5M, the envelope requirement also does not apply to features for which a straightness tolerance of the axis (even if smaller than the size tolerance) is specified.

The envelope requirement may also be indicated by a form tolerance with 0 Ⓜ applied (see 9.3.4). However, it is rather difficult to find the proper drawing indication (Figure 10.1). Therefore the symbol Ⓔ has been standardized in ISO 8015.

10.2 APPLICATION OF ENVELOPE REQUIREMENT

The envelope requirement is applicable to single features of size (cylindrical surfaces or features established by two parallel opposite plane surfaces).

The envelope requirement may be applied to features of size that are to be mated when the expected form deviations are larger than half of the least clearance or larger than permissible for interference fits in the maximum material condition.

Figure 10.2 shows the application of the envelope requirement for a cylindrical feature. The size tolerance requires that the actual local sizes are within the limits of size. In addition, the envelope requirement is specified.

Positional Tolerancing

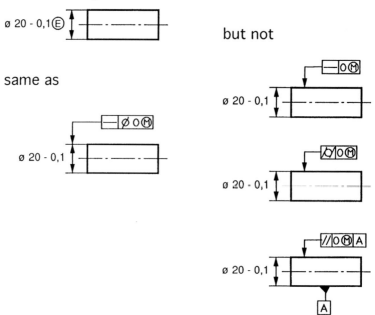

Figure 10.1: Drawing indication for the envelope requirement.

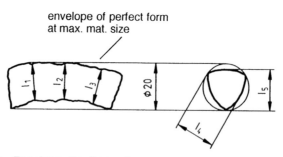

Figure 10.2: Envelope requirement.

126

When the envelope requirement is not specified, the actual local sizes within the cross-section could be within the limits of size; however, the cross-section could have a lobed form and go beyond the circle of maximum material size by the amount of the roundness tolerance (e.g. general roundness tolerance, see 16) (Figure 10.3). In addition, the feature could go beyond the envelope of perfect form at maximum material size by the amount of the straightness tolerance (e.g. general straightness tolerance) even if the feature is everywhere at maximum material size (Figure 10.4). In other words, the feature may violate the envelope requirement even when the size tolerance and form tolerance are respected. Therefore inspection of the size deviation and of the form deviation alone is not sufficient to verify the envelope requirement.

In order to verify the envelope requirement, gauging or a gauging simulation on a coordinate measuring machine must be executed (see 18.7.11), and/or the manufacturing tolerance must be diminished at the maximum material side by the amount of the expected form deviations.

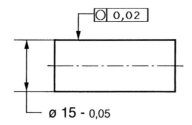

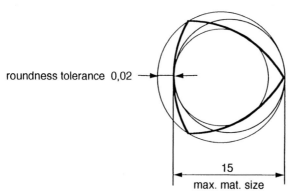

Figure 10.3: Violation of the envelope requirement even though size and roundness tolerances are respected.

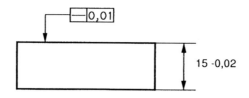

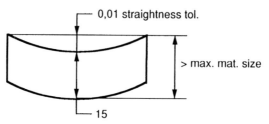

Figure 10.4: Violation of the envelope requirement even though size and straightness tolerances are respected.

10.3 CROSS-SECTIONS WITHIN SIZE TOLERANCE FIELDS

The envelope requirement deals with three-dimensional features. Sometimes, however, specifications deal with the circumference lines in cross-sections (e.g. in the German Standards DIN 1748 and DIN 17615 on aluminium sections). The specification requires the circumference lines in cross-sections normal to the axis to be contained between limiting lines that are geometrical ideal, concentric, in the geometrical ideal orientation relative to each other, and of distances equal to the maximum material sizes and the least material sizes (Figure 10.5).

This specification differs from the envelope requirement as follows:

• the entire feature is not considered, but only, cross-sections and each cross-section is taken to be independent of the others;

• not only is the geometric ideal form at maximum material size specified, but also that at least material size;

• the specification is not limited to single lines but to all lines establishing the entire circumference and taking into account the geometrical ideal orientation and location of the line elements relative to each other.

Symbols for this specification are not standardized.

The form deviations along the axis (straightness tolerance and flatness tolerance)

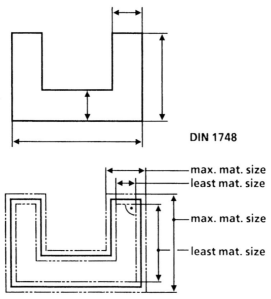

DIN 1748

max. mat. size
least mat. size

max. mat. size

least mat. size

Figure 10.5: Cross-sections between concentric geometrical ideal limiting lines of geometrical ideal orientation and of distances equal to the maximum material sizes and least material sizes.

are to be specified separately. DIN 1748 and DIN 17 615 specify the twist tolerance (see 3.7), instead of a flatness tolerance.

11 Least Material Requirement

11.1 DEFINITIONS

Least material condition (LMC): This is the state of the considered feature in which the feature is everywhere at that limit of size where the material of the feature is at its minimum, e.g. the maximum limit of size of a hole (maximum hole diameter) and the minimum limit of size of a shaft (minimum shaft diameter) (ISO 2692).
The actual axis of the feature need not be straight.

Least material size (LMS): This is the dimension defining the least material condition of a feature (ISO 2692), i.e. the limit of size where the material is at the minimum (e.g. the maximum limit of size of a hole and the minimum limit of size of a shaft).

Least material virtual size (LMVS): This is the size generated by the collective effect (with regard to what can be cut out of the material) of the least material size and the geometrical tolerance followed by the symbol Ⓛ, i.e.

for shafts: LMVS = LMS − geometrical tolerance;
for holes: LMVS = LMS + geometrical tolerance.

Least material virtual condition (LMVC): This comprises the features limiting the boundary of perfect (geometrical ideal) form and of least material virtual size. When more than one feature or one or more datum features are applied to the geometrical tolerance, the LMVS are in theoretical exact locations and orientations relative to each other (ISO 2692).

11.2 DESCRIPTION OF LEAST MATERIAL REQUIREMENT

Least material requirement (LMR): This requirement, indicated on drawings by symbol Ⓛ placed after the geometrical tolerance of the toleranced feature or after the datum letter in the tolerance frame, specifies

* when applied to the toleranced feature, the least material virtual condition (LMVC) of the toleranced feature should be fully contained within the material of the actual toleranced feature;

• when applied to the datum, the least material virtual condition (LMVC) of the datum feature should be fully contained within the material of the actual datum feature (ISO 2692).

The least material virtual condition at the datum feature is usually at least material size, because usually at the datum feature no geometrical tolerance followed by the symbol Ⓛ is indicated.

The least material requirement (LMR) Ⓛ has the effect that the least material virtual condition is entirely contained in the material of the feature and, for example, can be cut out. The mutual dependence of size and form or location and orientation is thereby taken into consideration.

11.3 APPLICATION OF LEAST MATERIAL REQUIREMENT

11.3.1 General

The least material requirement can be applied only to those features that have an axis or median plane (cylindrical features or features composed of two parallel opposite planes) (Table 9.1).

The least material requirement may be applied where a minimum material thickness must be respected and this minimum thickness depends on the mutual effect of deviations of size and form, location or orientation.

11.3.2 Least material requirement for the toleranced feature

The least material requirement for the toleranced feature allows an increase in the geometrical tolerance when the feature deviates from its least material condition (in the direction of the maximum material condition), provided that the least material virtual condition is not violated (is entirely within the material) (Figures 11.1 and 11.2). That is, the least material requirement specifies that the indicated geometrical tolerance applies when the feature is in its least material condition (smallest shaft and largest hole). When the feature deviates from the least material condition (larger shaft and smaller hole), the geometrical deviation may be larger without violating the least material thickness.

Figure 11.1 shows an example where a minimum ridge thickness should be contained in the material, e.g. of a casting that is to be machined. Figure 11.2 shows a permissible workpiece. The least material virtual condition has the size of 50 (least material size minus positional tolerance), is perpendicular to the datum (surface) A and 60 apart form the datum (surface) B. The more the ridge thickness deviates from the least material size, the more the actual median face may deviate from the theoretical exact location. In the example the smallest actual local size is 53. In this case the positional deviation of the actual median face may be 1.5, which corresponds to a positional tolerance of 3 (see Table 18.1).

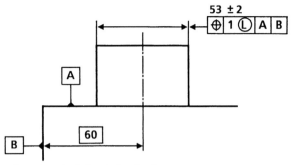

Figure 11.1: Least material requirement.

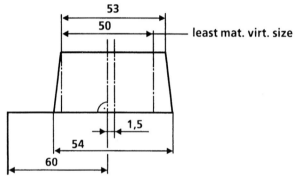

Figure 11.2: Permissible workpiece according to the tolerancing in Figure 11.1

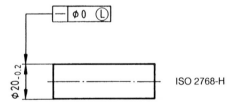

Figure 11.3: Straightness tolerance of the axis with least material requirement.

Figure 11.3 shows an example where the geometrical ideal form of least material size should be contained in the material (must be possible to cut out). Here the least material virtual condition has least material size.

Without this indication, the feature may be bent and may have lobed form in the cross-sections (here within the general tolerances of straightness and roundness according to ISO 2768-H). In the extreme most unfavourable case the maximum

Least Material Requirement

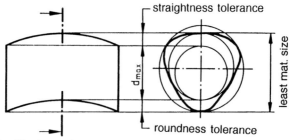

straightness tolerance

d_{max}

least mat. size

roundness tolerance

Figure 11.4: Maximum inscribed cylinder of diameter d_{max} in a workpiece of maximum permissible straightness deviations and with lobed forms of least material size and maximum permissible roundness deviations.

(geometrical ideal) cylinder contained in the material (possible to cut out) has a diameter of least material size minus the sum of straightness tolerance and roundness tolerance (Figure 11.4). The indication according to Figure 11.3 exludes this diminishing of the maximum contained cylinder.

11.3.3 Least material requirement for the datum

The least material requirement for the datum permits floating of the datum axis or datum median plane relative to the toleranced feature(s) when the datum feature deviates from its least material condition (in the direction of the maximum material condition). A prerequisite is that the surface of the datum feature does not violate the least material virtual condition (which is geometrically ideally positioned in relation to the geometrical ideal position of the toleranced features). Around this boundary, the datum feature may take, if possible, the position where the requirements of the toleranced features are fulfilled.

It should be noted that the indication of the least material requirement Ⓛ at the datum has a meaning different from ISO 5459. With A Ⓛ the datum cylinder A of least material size must be contained in the material (and is not allowed to lie out of the material (as applicable for the maximum material virtual condition of A Ⓜ).

Figure 11.5 shows an example where a geometrical ideal pipe with a minimum wall thickness of 5 is to be contained in the material (must be possible to cut out). Figure 11.6 shows the coaxial least material virtual cylinders that must not be violated by the surfaces. On the left-hand side both surfaces (features) have least material size, and therefore should be coaxial. On the right-hand side are shown the extreme permissible coaxiality deviations with maximum material sizes.

See also 20.10.4.

134

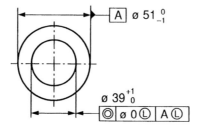

Figure 11.5: Coaxiality tolerance with least material requirement in order to secure the minimum wall thickness.

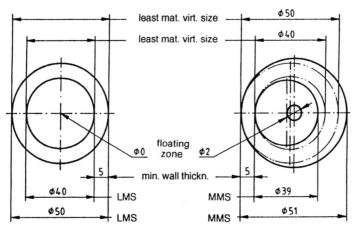

Figure 11.6: Calculation of the minimum wall thickness 5 according to tolerancing according to Figure 11.5: left with least material sizes; right with maximum material sizes.

11.3.4 Reciprocity requirement associated with least material requirement

Figure 11.7 shows three rings (a), (b) and (c). In all cases the same least material virtual conditions (boundaries to be entirely contained within the material) apply. In case (a) the whole tolerance is indicated at the size. The tolerance may be utilized by size deviations and by coaxiality deviations in an arbitrary, way. In case (b) the tolerance is distributed on size and coaxiality. The coaxiality tolerance will be enlarged by the size tolerance not utilized, but not vice versa. The size tolerance cannot be enlarged by the non-utilized coaxiality tolerance. In case (c) the size tolerance can also be enlarged by the non-utilized coaxiality tolerance.

Least Material Requirement

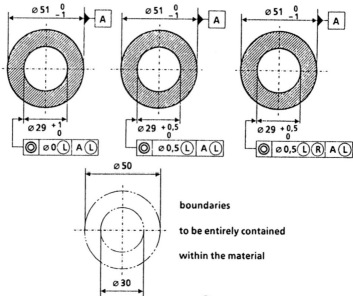

Figure 11.7: Least material requirement ($\textcircled{L}$ after geometrical tolerance and after datum letter) and reciprocity requirement ($\textcircled{R}$ after $\textcircled{L}$ after geometrical tolerance).

Reciprocity requirement RR associated with the least material requirement: This requirement, indicated on drawings by the symbol $\textcircled{R}$ placed after the symbol $\textcircled{L}$ after the geometrical tolerance in the tolerance frame, specifies that the least material virtual condition (LMVC) of the toleranced feature should not be violated. Deviations of size, form, orientation and location may take full advantage of the total tolerance (sum of tolerances).

The reciprocity requirement has the same effect as $0\,\textcircled{L}$, i.e. the total tolerance may be utilized for deviations of size, form, orientation or location in an arbitrary way. However, in contrast to $0\,\textcircled{L}$, the drawing indication with the reciprocity requirement $\textcircled{R}$ gives a recommendation to the manufacturer for the distribution of the total tolerance on size and geometrical characteristics. Thus the reciprocity requirement provides a communication between production planning and workshop.

It was discussed whether the reciprocity requirement should be symbolized with $\textcircled{R}$ after the size tolerance (because it modifies the latter) or with $\textcircled{M}$ or $\textcircled{L}$ after the size tolerance (in order to avoid the introduction of a new symbol) or by $\textcircled{R}$ after the symbol $\textcircled{M}$ or $\textcircled{L}$ after the geometrical tolerance in the tolerance frame. The Draft International Standard ISO DIS 2692 provides an Annex with the latter.

12 Tolerancing of Flexible Parts

Certain flexible or non-rigid parts, made for example from thin sheet metal, fibreglass, plastics or rubber, when removed from their manufacturing environment, deform substantial from their manufactured condition (geometrical shape) by virtue of their weight or flexibility, or by the release of internal stresses resulting from the manufacturing process.

When in the drawing in or near the title box "ISO 10 579 - NR" is indicated, all geometrical tolerances not associated with the symbol(F) apply in the restrained (assembled) condition. The restrained (assembled) condition is to be defined on the drawing (see e.g. Figures 12.1, 12.2 and 20.115).

Geometrical tolerances followed by the symbol (F) apply in the free state.

For the free state the conditions should be indicated under which the geometrical tolerance under this state is ensured (e.g. the direction of gravity) and, if necessary, the setting condition (support) of the part.

According to ISO 10 579, for the restrained condition, only those pressures and forces may be applied that can be expected under normal assembly conditions.

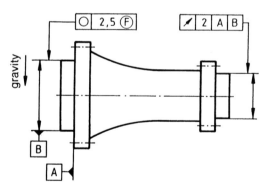

Figure 12.1: Tolerancing of a non-rigid (flexible) part according to ISO 10 579. ISO 10 579 - NR Restrained condition: Surface A is mounted with 64 bolts M6x1, screwed with a torque of 9 to 15 N m and the B feature is restrained to the corresponding maximum material limit.

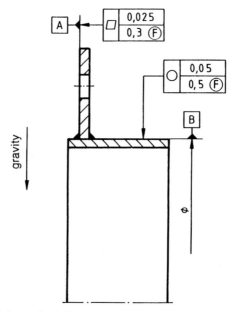

Figure 12.2: Tolerancing of a non-rigid (flexible) part according to ISO 10 579. ISO 10 579 - NR Restrained condition: Surface A is mounted with 120 bolts M 20, screwed with a torque of 18 to 20 N and the B feature is restrained to the corresponding maximum material limit.

See also 20.11.

13 Tolerance Chains (Accumulation of Tolerances)

When workpieces are fitted together, size deviations accumulate. In order to assess the resulting clearance or interference tolerance, tolerance line-up calculations are performed. Arithmetical tolerance calculations are based on extreme cases when all sizes are at their favourable or unfavourable limit of size. Statistical tolerance calculations take into acount the form of distribution of the local sizes and give the clearance or interference that will not be exceeded with a certain statistical probability (see 14).

The procedure for an arithmetical tolerance line-up calculation is as follows:

(1) define the dimension scheme, showing all dimensions and their tolerances that form the chain, i.e. all dimensions that contribute to the clearance or interference;

(2) dimensions whose upper limits lead to an increase in the closing dimension (clearance) are drawn in the positive direction, and others in the negative direction;

(3) the arithmetical sum of the maximum limits of sizes of the positive chain links and the minimum limits of size of the negative chain links gives the **maximum value of the closing dimension** (maximum clearance and minimum interference).

The arithmetical sum of the minimum limits of size of the positive chain links and the maximum limits of size of the negative chain links gives the **minimum value of the closing dimension** (minimum clearance and maximum interference). Instead of the limits of size, the permissible deviations may be taken when the arithmetical sum of the nominal sizes is zero (Figure 13.1).

The tolerance line-up calculation according to Figure 13.1 considers size tolerances only. It is supposed that the workpiece surfaces respect the geometrical ideal boundaries of maximum material size. In the example of Figure 13.2 this is the boundary consisting of a cylinder of minimum size of the hole and of two parallel planes the minimum distance apart and perpendicular to the cylinder.

This supposition is not made in ISO 8015. Therefore the parts must be toleranced appropriately (Figure 13.2).

The tolerancing of the pependicularity deviation (to be within the size tolerance) may also be chosen according to Figure 13.3. The meaning (i.e. the go gauge) of

139

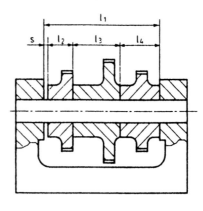

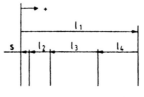

$$s_{max} = l_{1max} - l_{4min} - l_{3min} - l_{2min}$$

$$s_{min} = l_{1min} - l_{4max} - l_{3max} - l_{2max}$$

Figure 13.1: Arithmetical tolerance calculation.

both Figures 13.2 and 13.3 is the same. In Figure 13.3 the indication "A—B Ⓜ" has the same effect as the indication "common zone" in Figure 13.2.

When the perpendicularity deviations are not toleranced by "0 Ⓜ", the workpieces may have deviations as shown in Figure 13.5.

The tolerance calculation as described in Figure 13.1 is then not sufficient. The perpendicularity deviation may override the minimum clearance, and the assembly may jam. The tolerance calculation must then include the geometrical tolerances (Figure 13.6).

It should be noted that the perpendicularity tolerance t_{r1} of the faces of the dimension l_1 decreases the clearance, and must therefore enter the dimension scheme in the negative direction (this in opposition to the dimension l_1 itself) (Figure 13.8).

When the perpendicularity tolerances are smaller than the general tolerances, they must be indicated in the drawing (e.g. according to Figures 13.4 and 13.9). Because this is an assembly with clearance, the maximum material requirement is appropriate to the function. It allows larger perpendicularity deviations when the workpiece sizes deviate from the maximum material size (approach the minimum material size).

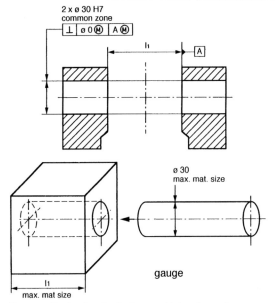

Figure 13.2: Perpendicularity deviations within size tolerances.

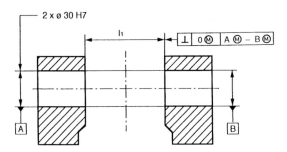

Figure 13.3: Perpendicularity deviations within size tolerances.

When the perpendicularity tolerances are larger than the general tolerances, the latter are sufficient, and the indication of the perpendicularity tolerances and of the maximum material requirement can be omitted (but have to be taken into account in the tolerance line-up calculations).

Figure 13.10 shows an assembly and the tolerance line-up calculation, taking into account the perpendicularity tolerances of the side faces. In this case the assembly may jam although the tolerances are respected, because the right face has a smaller diameter than the adjacent part (Figure 13.11). Tolerancing taking account of the size of the counterpart is appropriate to the function (Figure 13.12).

Tolerance Chains (Accumulation of Tolerances)

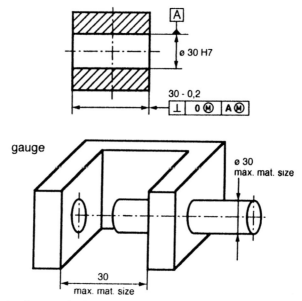

Figure 13.4: Perpendicularity deviations within size tolerances.

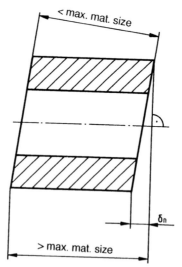

Figure 13.5: Workpiece with maximum material sizes and perpendicularity deviations.

Tolerance Chains (Accumulation of Tolerances)

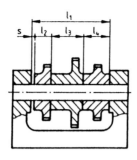

$$S_{max} = l_{1max} - l_{4min} - l_{3min} - l_{2min}$$
$$S_{min} = l_{1min} - t_{r1} - l_{4max} - t_{r4} - l_{3max} - t_{r3} - l_{2max} - t_{r2}$$

Figure 13.6: Arithmetical tolerance calculation including the perpendicularity tolerances t_r $\textcircled{M}$.

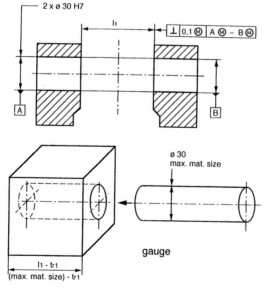

gauge

Figure 13.7: Assembly part with perpendicularity tolerance indicated.

143

Tolerance Chains (Accumulation of Tolerances)

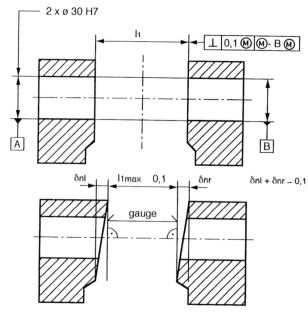

Figure 13.8: Clearance-decreasing effect of perpendicularity deviation.

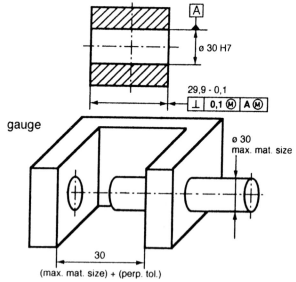

Figure 13.9: Assembly part with perpendicularity tolerance indicated.

Tolerance Chains (Accumulation of Tolerances)

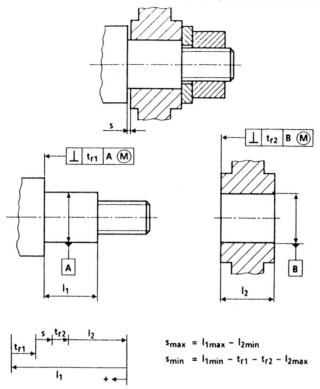

$$s_{max} = l_{1max} - l_{2min}$$
$$s_{min} = l_{1min} - t_{r1} - t_{r2} - l_{2max}$$

Figure 13.10: Tolerancing and tolerance calculation taking into account the perpendicularity tolerances t_r.

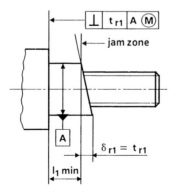

Figure 13.11: Permissible position of the side face.

Tolerance Chains (Accumulation of Tolerances)

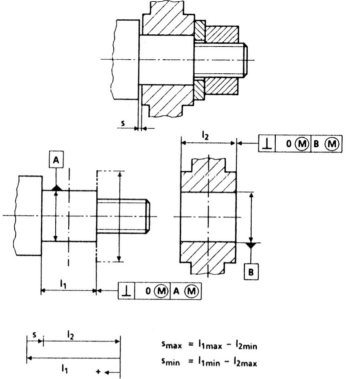

$$s_{max} = l_{1max} - l_{2min}$$
$$s_{min} = l_{1min} - l_{2max}$$

Figure 13.12: Tolerancing and tolerance calculation taking account of the size of the counterpart.

14 Statistical Tolerancing

With the arithmetical tolerance line-up calculation, the tolerances of the links of the dimension chain (e.g. single parts of an assembly) are defined in such a way that the assembly still functions when all links of the chain have actual sizes equal to their size limits (Figure 14.1).

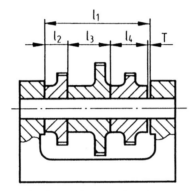

Figure 14.1: Arithmetical tolerancing, single tolerances T_n and closing tolerance $T = T_1 + T_2 + ... + T_n$.

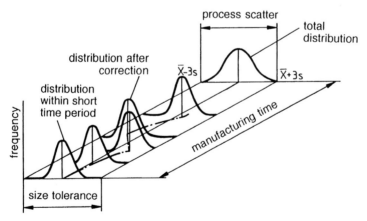

Figure 14.2: Distribution of actual sizes during a manufacturing process.

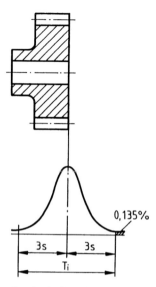

Figure 14.3: Distribution of actual sizes (size deviations) of a part.

However, in practice the actual sizes are subject to variations, and are therefore statistically distributed (e.g. according to a normal distribution) (Figures 14.2 and 14.3). With this supposition, it is most unlikely that an assembly contains only parts with actual sizes equal to the size limits (Figure 14.4).

In contrast to the arithmetical tolerance line-up calculation, in the statistical tolerance line-up calculation it is assumed that this extreme case will not occur. The single tolerances are chosen larger than with arithmetical tolerancing.

The functional closing tolerance T of the assembly is divided up into n single tolerances with arithmetical tolerancing according to

$$T = T_a = T_1 + T_2 + ... + T_n,$$

with statistical tolerancing according to

$$T = T_s = \sqrt{(T_{s1}^2 + T_{s2}^2 + ... + T_{sn}^2)}.$$

With an assembly chain of four parts (Figure 14.5), the following applies.

With arithmetical tolerancing, the single tolerance is $T/4 = 0.25T$. With statistical tolerancing, the single tolerance is $T/\sqrt{4} = 0.5T$. Non-functioning occurs when the arithmetical sum of the actual deviations exceeds the closing tolerance. The probability for this is, with statistical tolerancing (variation of the single dimension equal to one sixth of the statistical tolerance and variation of the closing dimension

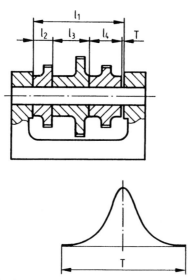

Figure 14.4: Distribution of actual values of the closing dimension.

equal to one-sixth of the closing tolerance),

$$P = 0.135\%,$$

i.e. 1.35 of 1000 assemblies may jam, and at least one part must be changed for another one.

Here it is assumed that the actual sizes are normally distributed, their standard deviation is one-sixth of the statistical single tolerance, the distribution is centred in the tolerance and the (dimensions of the) parts are independent of each other.

Deviations from these assumptions are covered by introducing a safety factor c:

$$T_s = c\sqrt{\sum T_{si}^2}.$$

Depending on the manufacturing method (form of distribution) and according to the experiences of the manufacturer, the safety factor c ranges between 1.4 and 1.8, mostly 1.5.

In Ref. [2] safety factors c are indicated, depending on the form of distribution of the single sizes (equal, trapeze, triangular, normal). According to the central limit theorem of mathematical statistics, in any case, independently of the kind of distribution of the single sizes, the closing dimensions are approximately normally distributed if more than four members contribute to the closing dimension.

In order to stipulate and to verify the assumptions of statistical tolerancing, properly defined terms are needed. An International standard on this subject does

Statistical Tolerancing

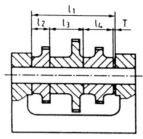

Arithmetical tolerancing: single tolerance   Tai = T / 4   = 0,25 T
Statistical tolerancing: single tolerance   Tsi = T /√4   = 0,5 T

single dimensions

convolution
of 4 distributions
(sizes)

closing dimension
(clearance)

0.135%

$3s_{si}$ | $3s_{si}$

T_{si}
= 0.5 T

0,135%

$3s_T$ | $3s_T$

T

Figure 14.5: Probability of exceeding the statistical closing tolerance of a chain of four parts. Probability of exceeding T: 0.135%, i.e. 1.35 of 1000 assemblies.

not yet exist. The German Standard DIN 7186 defines terms as follows:

Mean size C   This is the arithmetical mean of the limits of size.

Dimension chain   This is a geometrical representation of several coeffecting dimensions that are independent of each other (Figure 14.6).

Single dimension M   This is a dimension acting as a link in a dimension chain.

Closing dimension M_o   This is the result of arithmetic addition of n independent single dimension within a dimension chain.

Single tolerance T_i   This is the tolerance of a single dimension.

150

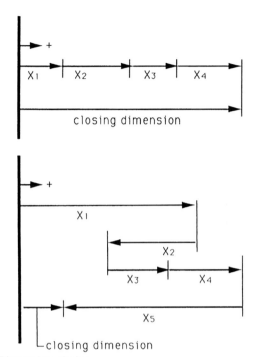

Figure 14.6: Dimension chains.

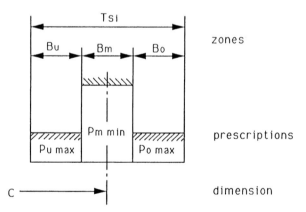

Figure 14.7: Statistical tolerancing, tolerance zones. B_u, B_o lower and upper side zones; B_m central zone; P_{umax}, P_{omax} maximum lower, maximum upper side content; P_{mmin} minimum central content; C mean size.

Statistical Tolerancing

Statistical tolerance T_{si} This is a single tolerance with specifications regarding the distribution of the actual sizes within the tolerance zone.

Arithmetical closing tolerance T_a This is the sum of the single tolerances within a dimension chain.

Square closing tolerance T_q This is the positive root of the sum of squares of the single tolerances wihin a dimension chain. It is the least possible statistical tolerance.

Statistical closing tolerance T_s This is a closing tolerance, smaller than T_a, specified according to the actual size distribution.

Side zones and central zone For the specification of statistical tolerances the tolerance zone is divided in zones. In general, three zones predominently symmetrical with respect to the mean size C are sufficient (Figure 14.7).

Lower side zone B_u This is the zone adjacent to the minimum limit of size.

Upper side zone B_o This is the zone adjacent to the maximum limit of size.

Central zone B_m This is the range (zone) between the upper and lower side zones.

Side contents P_u and P_o These are the percentages of actual sizes (more precise: mating size) of a manufacturing lot within the side zone (Figure 14.7). If not otherwise specified, according to DIN 7186 the contents P_u and P_o must not be more than $(100\% - P_m\%)/2$ each. (From this it follows that, with an asymmetrical distribution of the actual sizes of a manufacturing lot, much more than the specified $P_m\%$ must be contained in the centre zone B_m.)

Central content P_m This is the percentage of actual sizes (more precise: mating size) of a manufacturing lot within the central zone (Figure 14.7).

The following specifications are predominant:

For **machining processes (metal removal)**

- central zone (width) and side zones (widths) each one-third of the tolerance T_{si}

- central content at least 50%, side contents not more than 25% each.

For **bending and punching processes**

- central zone $0.5T_{si}$, side zones $0.25T_{si}$ each;
- central content at least 50%, side contents not more than 25% each.

Figure 14.8 shows the drawing indication according to DIN 7186.

Statistical tolerancing allows specifications of larger single tolerances than arithmetical tolerancing without detriment to the function of the workpiece. In many cases this results in a gain in manufacturing economy. Because of the current trend to greater miniaturization and to more precise products, statistical tolerancing has become more important. Sometimes smaller arithmetical tolerances are not even achievable.

However, the following prerequisites should be for manufacturing:

- the actual sizes of a single dimension are approximately normally distributed;
- the mean values of the distributions coincide approximately with the mean size;
- the ratios of statistical tolerance T_{si} and standard deviation σ are of the same order of magnitude, e.g. $T_{si}/\sigma = 6$;
- there is no mutual dependence between the dimensions within the dimension chain.

Statistical tolerancing is more advantageous

- the greater the number of menbers in the dimension chain;
- the greater the manufacturing lot;
- the better manufacturing and inspection can satisfy the prerequisity for statistical tolerancing.

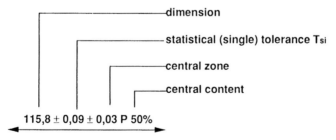

Figure 14.8: Statistical tolerancing; drawing indication and interpretation according to DIN 7186.

153

The above apply when only size tolerances are considered. When geometrical tolerances must be considered it is necessary to take into account that geometrical deviations in general are not normally distributed. Their distribution more closely resembles a logarithmic one.

A practically used method to cope with geometrical tolerances for statistical tolerancing is to compare the logarithmic distribution with the centre zone and one side zone (Figure 14.9). Then the mean size C is 25% and the centre zone $\pm$ 25% of the geometrical tolerance.

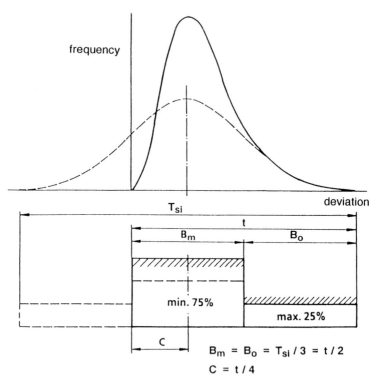

$$B_m = B_o = T_{si} / 3 = t / 2$$
$$C = t / 4$$

drawing indication:

$$t / 4 \,^{+\,3\,t\,/\,4}_{-\,t\,/\,4} \pm t / 4 \; P75\%$$

Figure 14.9: Statistical tolerancing with an unsymmetrical distribution of geometrical deviations.

Note that this mean size C should be distinguished from the desired value 0 for the statistical process control SPC and from the target value 0 of the Taguchi method.

For the statistical tolerance line-up calculation the virtual tolerance T_{si} that covers the virtual lower side zone (Figure 14.9) has to be taken into account.

Figure 14.10 shows an example: with the drawing indication at the bottom and the interpretation at the top.

Figure 14.11 gives a comparison of arithmetical and statistical tolerancing. In this example the tolerances for linear dimensions are more than twice as large with statistical tolerancing as with arithmetical tolerancing, and the tolerances for perpendicularity are 50% larger, with no expected detriment to the function.

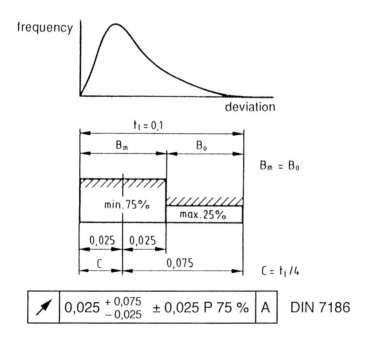

Figure 14.10: Statistical tolerancing; example of run-out tolerance.

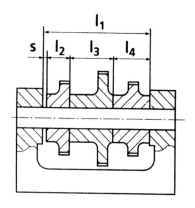

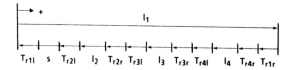

$n = 12$ (4 lengths, 8 perpendicularity tolerances)

$s_{min} = 0,1$ $s_{max} = 0,6$ $\Delta s = 0,5$

Arithmetical tolerancing:

$T_{ai} = \Delta s / n$ $= 0,5 / 12$ $= 0,042$

single dimensions, e.g.: $20 \pm 0,021$
perpendicularity tolerances: $0,042$

Statistical tolerancing:

$T_{si} = \Delta s / c \sqrt{n}$ $= 0,5 / 1,5 \sqrt{12}$ $= 0,096$

single dimensions, e.g.: $20 \pm 0,048 \pm 0,016$ P50%

perpendicularity tolerances: $0,016 \begin{smallmatrix} +\,0,048 \\ -\,0,016 \end{smallmatrix} \pm 0,016$ P75%

Figure 14.11:   Comparison of arithmetical and statistical tolerancing.

15 Respecting Geometrical Tolerances during Manufacturing

15.1 MANUFACTURING INFLUENCES

Geometrical deviations are influenced by the following, which are sometimes referred to as the "5 M".

Material

- rigidity of the workpiece (shape);
- material;
- stress in the material.

Machine (tool)

- precision of the machine tool, bearing play;
- static and dynamic rigidity of the machine tool;
- thermal properties of the machine tool;
- maintainance;
- environment (e.g. vibrations).

Method

- tool;
- chuck, fixing, clamping method;
- processing data (e.g. cutting speed, thickness of cut); cutting pressure.

Measuring

- uncorrected systematic measuring deviations;
- random measuring deviations.

Manufacturer

- education, skillness, presision of re-chucking;
- environment.

The following figures show some results of investigations in the field of metal-removal processes. These results are only examples, and are not general.

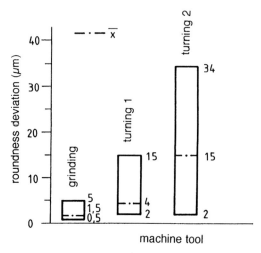

Figure 15.1: Ability of machine tool, mean values and variations of roundness deviations of workpieces made out of steel ∅ 20 mm.

Figure 15.1 shows the mean values $\bar{x}$ and the variations of measured roundness deviations of workpieces made out of steel, ∅ 20 mm, manufactured with certain machine tools.

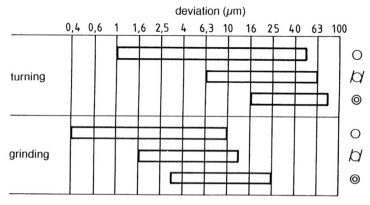

Figure 15.2: Ability of manufacturing devices in a particular workshop.

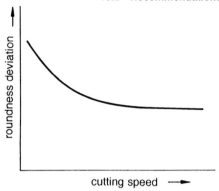

Figure 15.3: Roundness deviation as a function of cutting speed.

Figure 15.2 shows the ability of manufacturing devices to respect tolerances of roundness, cylindricity and coaxiality by turning and grinding.

The machine tools are classified within the variations. Their abilities related to workpiece material and shape and process data (e.g. cutting speed, thickness of cut and chucking method) are recorded for manufacturing planning.

Figure 15.3 shows, as an example, the typical dependence of roundness deviation on cutting speed

Figure 15.4 shows the result of chucking and cutting pressure influence on the workpiece form.

The largest geometrical deviations occur in general as a result of re-chucking the workpiece. Therefore special care should be taken in re-chucking.

15.2 RECOMMENDATIONS FOR MANUFACTURING

In order that **geometrical tolerances** are respected, the abilities of the machine tools should be assessed and be classified. In order that narrow geometrical tolerances are respected by the classification the process data (e.g. cutting speed) and workpiece properties (e.g. wall thickness) must be taken into account. Because re-chucking influences the geometrical deviations greatly, it should be specified in the manufacturing planning whether, it is to be re-chucked, and if so then, when, where, how and with what precision.

With regard to respecting **general geometrical tolerances** see 16.

In order that the **envelope requirement** Ⓔ is respected, it is recommended that the tolerance specified in the drawing at the maximum material limit be reduced by the expected form deviation, i.e. the actual sizes must differ from the maximum material size at least by the amount of the expected form deviation.*

---

*If appropriate, a manufacturing drawing should be issued using the reciprocity requirement (see 9.3.6, 11.4 and 20.11).

Cause   Effect

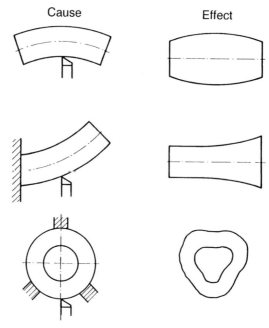

Figure 15.4: Influence of chucking and cutting pressure on the workpiece form.

In order that the **maximum material requirement** Ⓜ at the toleranced feature is respected, the following are recommended.

With drawing indication 0 Ⓜ   The same applies as for Ⓔ*†

With drawing indication t Ⓜ *(where t > 0)*   When this tolerance *t* is larger than the expected geometrical deviation, it may be manufactured as if Ⓜ were not specified. Only when the specified geometrical tolerance is exceeded is it necessary to check according to the maximum material requirement.†

When the indicated tolerance *t* is not larger than the expected geometrical deviation, it is recommended that the size tolerance at the maximum material limit be reduced accordingly.*†

---

†Features related by relatively small tolerances of orientation or location (toleranced feature(s) and datum feature(s)) should be manufactured without re-chucking

In order to respect the **maximum material requirement Ⓜ at the datum feature,** the following are recommended.

- When at the datum letter in the tolerance frame the symbol Ⓜ is indicated but the datum feature itself has no geometrical tolerance with specified Ⓜ, the same applies as for Ⓔ*†

- When at the datum feature itself a geometrical tolerance with specified Ⓜ is indicated, proceed in the same way as with *t* Ⓜ (where *t* > 0) at toleranced features given above.*†

The ISO Standards assume that the toleranced axis is an irregular line composed of the centres of cross-sections (see 3.5 and 18.7.2.1). In this case the workpiece surface respects the maximum material condition when the actual axis is within the orientation or location tolerance.

However, some national standards assume that the toleranced axis is a straight line. In these cases the workpiece surface may exceed the maximum material virtual condition although the "straight actual axis" is within the orientation or location tolerance. This because of the form deviations of the workpiece surface, which are eliminated by the assessment of the straight actual axis. In these cases the symbol Ⓜ must not be ignored as described above.

In order to respect the **least material requirement** Ⓛ, proceed in the same way as with the maximum material requirement, but reduce the size tolerance at the least material limit.

In order to respect the **projected tolerance zone** Ⓟ, it is recommended that the positional tolerance be reduced by twice of the expected angularity deviation related to the projected length (Figure 15.5).

In order to respect the **reciprocity requirement** Ⓡ, the manufacturing may be as if Ⓡ were not specified.

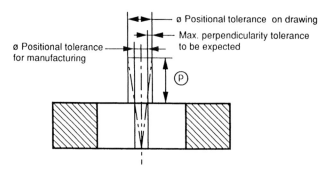

Figure 15.5: Recommendation for manufacturing in order to respect the projected tolerance zone Ⓟ

Figure 15.6 gives a synopsis of these recommendations. However, for the decision as to whether the workpiece meets the drawing specifications the definitions of the requirements must be observed.

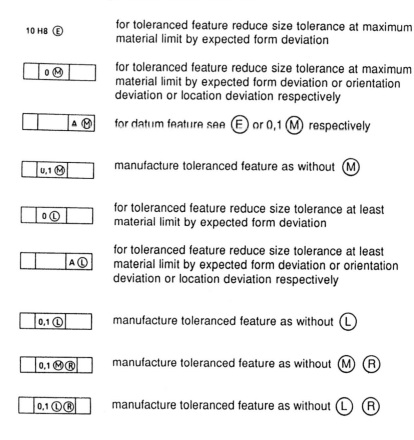

10 H8 Ⓔ — for toleranced feature reduce size tolerance at maximum material limit by expected form deviation

| 0 Ⓜ | — for toleranced feature reduce size tolerance at maximum material limit by expected form deviation or orientation deviation or location deviation respectively

| A Ⓜ | — for datum feature see Ⓔ or 0,1 Ⓜ respectively

| 0,1 Ⓜ | — manufacture toleranced feature as without Ⓜ

| 0 Ⓛ | — for toleranced feature reduce size tolerance at least material limit by expected form deviation

| A Ⓛ | — for toleranced feature reduce size tolerance at least material limit by expected form deviation or orientation deviation or location deviation respectively

| 0,1 Ⓛ | — manufacture toleranced feature as without Ⓛ

| 0,1 ⓂⓇ | — manufacture toleranced feature as without Ⓜ Ⓡ

| 0,1 ⓁⓇ | — manufacture toleranced feature as without Ⓛ Ⓡ

Features which are related by relatively small tolerances of orientation or location (toleranced feature(s) and datum feature(s)) should be manufactured without re-chucking

Figure 15.6: Recommendations for manufacturing in order to respect Ⓔ, Ⓜ, Ⓛ and Ⓡ.

16 General Geometrical Tolerances

16.1 DEMAND FOR GENERAL GEOMETRICAL TOLERANCES

As the principle of independency demands an indication for each requirement, it calls for the application of a standard on general geometrical tolerances (title block tolerances on geometry). Otherwise the drawing would be embroidered with geometrical tolerance indications (Figure 16.11).

Even if Rule #1 of ANSI Y14.5M (see 21.2) is used, there is still a need for general geometrical tolerances on orientation (perpendicularity) and location (coaxiality and symmetry), because Rule #1 does not apply to related features. Furthermore, there is a need for general geometrical tolerances on form for single features to which a disclaimer from Rule #1 is indicated.

16.2 THEORY OF GENERAL TOLERANCES

Before applying general tolerances, their theory must be agreed upon. This theory has been developed in the International Organization for Standardization ISO as follows (ISO 2768):

(a) Each feature requires limits on its deviations determined by its function.

(b) The drawing must be definitive, i.e. it must specify all dimensional and geometrical tolerances necessary to completely define the shape and size of the part.

(c) Above certain values of tolerances, there is generally no gain in economy by increasing them further. These tolerances are not exceeded in normal workshop practice without particular effort. This is the normal (customary) workshop accuracy.

(d) General tolerances take account of normal workshop accuracy. They are specified on the drawing by a general tolerance note, usually in or near the title block referring to a standard specifying the general tolerances (Figure 16.1).

(e) If the function requires tolerances smaller than the general ones, they must be indicated individually.

If the function allows tolerances equal to or greater than the general ones,

163

they should not be indicated individually. For the features concerned the general tolerances should apply.

(f) Exceptions to the rule are where the functions allow greater tolerances and these greater tolerances are more economical in the particular cases than the general tolerances (e.g. the length of blind holes drilled at assembly). In these few exceptional cases the tolerance required by the function, although greater than the general tolerance, should be indicated individually.

(g) Before the introduction of general tolerances into the workshop, the tolerance class (accuracy grade) of general tolerances to which its normal workshop accuracy corresponds should be assessed. (The general tolerances of that tolerance class should be reliably respected, i.e. they should be so large that they are well within the manufacturing capability of the ordinary machine tools.

(h) The workshop should only accept drawings where the general tolerance class is equal to or coarser than its normal workshop accuracy.

(i) The workshop should ensure by sampling inspection that its normal workshop accuracy is not impaired in the course of time.

(j) If a general tolerance is exceeded (although this is very unlikely, see (c) and (g)–(i)), the decision whether or not the part is to be rejected depends on the function (see (c) and (e)).

General tolerances are not guidelines that may be exceeded without risk, as has sometimes been wrongly supposed. The probability that exceeding them will lead to rejection of the part is merely less than with individually indicated tolerances.

(k) The use of general tolerances leads to the following advantages:

• drawings are easier to read (not being embroidered with tolerance indications) (Figure 16.11);

• the designer saves a great deal of detailed tolerance calculations (it is sufficient to know that the function allows tolerances = general tolerances);

• drawings indicate which characteristics can be produced with normal manufacturing capability (ordinary machine tools);

• drawings indicate which characteristics are not likely to be exceeded (checking level can be less frequent).

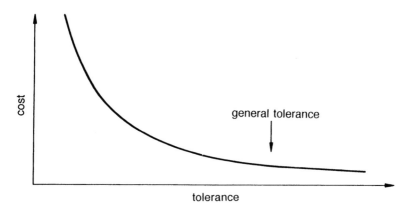

Figure 16.1: Cost function of tolerances; general tolerances.

Sometimes concepts are proposed that deviate from the above, for example for distinguishing between functional dimensions and non-functional dimensions. The tolerances for functional dimensions should then be indicated individually, whereas the general tolerances should apply only to the non-functional dimensions. As mentioned above, there are limits derived from the function for each dimension (feature). Therefore each dimension is a functional one, and it is impossible to distinguish between them under these circumstances. A similar argument applies to geometrical characteristics.

Other concepts propose applying the general tolerances only to those characteristics the functional tolerance for which is much greater than the general tolerance. But then the question arises as to by how much the functional tolerance should be greater than the general tolerance in order to apply the latter. If a factor is established (e.g. 2) then why not double the general tolerance?

An objection sometimes heard against general geometrical tolerances is that they would also control characteristics that are not essential for the function of the workpiece (see Figure 20.17). Since the general tolerances are so large that they are respected in any case (see (c)), there is no disadvantage in applying them to features that would allow wider tolerances.

What logically remains is the theory given above under (a)–(k).

16.3 DERIVATION AND APPLICATION OF GENERAL GEOMETRICAL TOLERANCES

The general geometrical tolerances for machined parts that are now standardized in ISO 2768-2 and that have been in use within parts of the industry for several years (e.g. in Germany) are described in the following.

165

These general geometrical tolerances, given in Tables 16.1–16.4, are derived from measurements on workpieces formed by metal removal. Mainly workpieces were measured to which the general dimensional tolerances ISO 2768-m had been applied (manufacturing of turbines, machine tools, fine mechanical engineering etc.). Only such features were measured that had no geometrical tolerance indication on the drawing.

16.3.1 Straightness and flatness

The general tolerances on straightness or flatness are given in Table 16.1. When selecting a straightness tolerance from the table, the length of the corresponding line must be taken, and in the case of a flatness tolerance the longer length of the surface or the diameter of the circular surface must be taken.

Table 16.1: General tolerances on straightness and flatness (values in mm)

Tolerance class	Straightness and flatness tolerances for ranges of nominal lengths					
	≤ 10	$> 10,$ ≤ 30	$> 30,$ ≤ 100	$> 100,$ ≤ 300	$> 300,$ ≤ 1000	$> 1000,$ ≤ 3000
H	0.02	0.05	0.1	0.2	0.3	0.4
K	0.05	0.1	0.2	0.4	0.6	0.8
L	0.1	0.2	0.4	0.8	1.2	1.6

The general tolerances given in Table 16.1 are derived from measurements as described above (Figure 16.2). Measured geometrical deviations have remained within the tolerances of class H in Table 16.1. This has been proved by measurements made in the industries of different countries (e.g. Germany, Japan and the UK).

According to measurements made in Switzerland, in some companies greater deviations occur. They remain within the tolerances of class K or L.

16.3.2 Roundness (circularity)

General tolerances on circularity have been established equal to the numerical value of the diameter tolerance (Figure 16.3) or to the respective value of the general tolerance on circular run-out in Table 16.4, whichever is the smaller.

The circularity deviation cannot be greater than the radial circular run-out deviation of the same feature. Therefore the numerical values of the general tolerances on circular run-out (Table 16.4) have been taken as the upper limits on the general tolerance s on circularity.

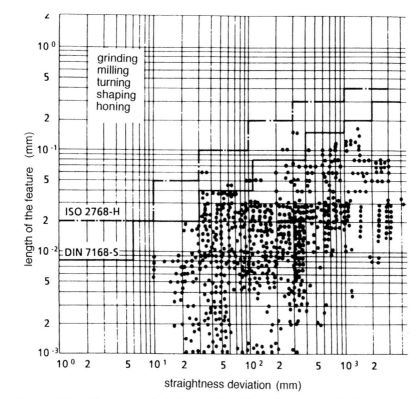

Figure 16.2:   Measured straightness deviations (measured in German and Japanese industry).

Whether the roundness deviation may take advantage of the full tolerance zone depends on the diameter tolerance and the shape of the roundness deviation.

In the case of an elliptic form the deviation may only occur up to half of the numerical value of the size tolerance, otherwise actual local sizes would exceed the size tolerance (Figure 16.4).

In the case of Figure 16.5 the form deviation may occur up to the size tolerance.

In the case of a three-lobed form the deviation can occur up to 0.15 times the maximum permissible diameter M_a (Figure 16.6). This might be more than the size tolerance, and could then be more than what is allowed by the general tolerance on circularity. Although such extreme form deviations seldom occur, those production methods in which lobed forms can appear should be observed in order not to exceed the general tolerances on circularity.

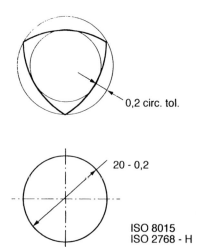

Figure 16.3: Roundness tolerance zone.

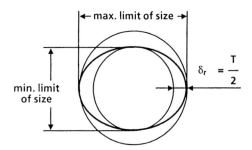

Figure 16.4: Maximum permissible elliptic form deviation.

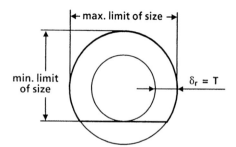

Figure 16.5: Maximum permissible form deviation.

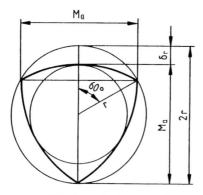

Figure 16.6: Maximum possible form deviation with three-lobed form.

16.3.3 Cylindricity

The cylindricity deviation consists of three components: circularity deviation, straightness deviation and parallelism deviation of opposite generator lines. Each of these is controlled by its general tolerance. However, it is very unlikely that on one workpiece all three components occur with their maximum permissible values and that they accumulate to the theoretical maximum permissible value of the cylindricity deviation. On the other hand, not enough measured values are presently available to derive suitable general tolerances on cylindricity.

Since the cylindricity deviation is almost only of importance for cylindrical fits, and since then it is already more closely limited by the envelope requirement Ⓔ, there seems to be no need for the establishment of general tolerances on cylindricity. Therefore they have been omitted.

16.3.4 Line profile, surface profile and position

Profile tolerances of lines and surfaces and positional tolerances necessitate the indication of theoretical exact dimensions (in rectangular frames) and therefore individual indications. Furthermore, these tolerances are usually relatively small, so that for this purpose general tolerances hardly come into question. Therefore they have been omitted. These characteristics have to be toleranced individually.

16.3.5 Parallelism

The limitation for parallelism deviations results from the straightness (or flatness) tolerance (Figure 16.7(b)) or from the size tolerance (Figure 16.7(a)), whichever is greater. Greater parallelism deviations are not to be expected, because otherwise either the straightness or flatness tolerance or the size tolerance would be exceeded.

169

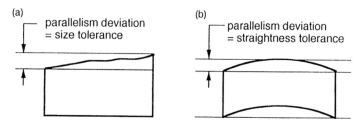

Figure 16.7: Parallelism deviation: (a) equal-size tolerance; (b) equal-straightness tolerance.

16.3.6   Perpendicularity

The general tolerances on perpendicularity are given in Table 16.2. The longer of the two sides forming the rectangular angle is to be taken as the datum. If the nominal lengths of the two sides are equal, either of them may apply as the datum.

In deriving the values in Table 16.2 the following has been observed. The definition of the geometrical tolerance on perpendicularity according to ISO 1101 as follows:

The zone between two parallel straight lines or planes or cylindrical zone that is perpendicular to the datum and within which the actual line or surface or the axis or median plane shall remain.

This tolerance zone also limits the straightness or flatness deviations and the axial run-out deviations of the sides forming the rectangular angle. Therefore the general tolerance on perpendicularity should not be smaller than the general tolerances on straightness (and flatness) and on axial run-out.

The general tolerances on perpendicularity given in Table 16.2 respect this, and are derived from measurements as described above. Measured deviations have remained within the tolerances of class H in Table 16.2.

Table 16.2:   General tolerances on perpendicularity (values in mm)

	Perpendicularity tolerances for ranges of nominal lengths of the shorter side			
Tolerance class	≤ 100	> 100, ≤ 300	> 300, ≤ 1000	> 1000, ≤ 3000
H	0.2	0.3	0.4	0.5
K	0.4	0.6	0.8	1
L	0.6	1	1.5	2

16.3.7   Angularity

General tolerances on angularity in the definition of ISO 1101 (the zone between 2 parallel lines or planes inclined to the datum in the theoretically exact angle indicated in a rectangular frame) are not specified. For angles indicated in angular dimensions the general tolerances according to ISO 2768-1 apply. For angles indicated in theoretical exact dimensions (e.g. 30°) the angularity tolerance has to be indicated on the drawing individually.

16.3.8   Coaxiality

In the extreme case the coaxiality deviation may be as great as the value given in Table 16.4 for the radial circular run-out tolerance, since the radial circular run-out deviation is composed of the coaxiality deviation and parts of the circularity deviation. General tolerances on coaxiality are not intended to be standardized.

16.3.9   Symmetry

The general tolerances on symmetry are given in Table 16.3. They apply to symmetrical features, also if one of the two features is symmetrical and the other cylindrical. The longer feature is to be taken as the datum. If the nominal lengths of the two features are equal, either of them may apply as the datum.

Table 16.3:   General tolerances on symmetry (values in mm)

Tolerance class	Symmetry tolerances for ranges of nominal lengths			
	≤ 100	> 100, ≤ 300	> 300, ≤ 1000	> 1000, ≤ 3000
H	0.5	0.5	0.5	0.5
K	0.6	0.6	0.8	1
L	0.6	1	1.5	2

In deriving the values in Table 16.3 the following has been observed. As the tolerance zone on symmetry also limits certain straightness or flatness deviations, the general tolerances on symmetry should not be smaller than the general tolerances on straightness and flatness. Futhermore, measurements on workpieces (as described above) revealed that symmetry deviations up to 0.5 mm occur independently of the feature length (Figure 16.8). The reason for this is probably the following. The general tolerances are set up depending on the greatest measured deviations. These deviations were not due to the inaccuracy of the machine tool but rather to inaccuracy when adjusting the workpiece in the machine tool after the workpiece has been turned over (re-chucking). Small and large workpieces were adjusted with the same inaccuracy, and showed the same distribution of the measured symmetry deviations. Greater symmetry deviations occur only owing to greater straightness or flatness deviations (classes K and L).

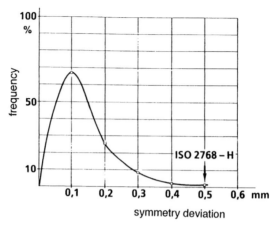

grinding
milling
shaping, planing
feature length < 500 mm

Figure 16.8: Measured symmetry deviations.

16.3.10 Circular run-out

The general tolerances on circular run-out* (radial, axial and inclined circular run-out) are given in Table 16.4. The bearing surfaces are to be taken as the datum if they are designated as bearing surfaces. In the other case, for radial circular run-out, the longer feature is to be taken as the datum. If the nominal lengths of the two features are equal, either of them may apply as the datum.

Table 16.4 is derived from measurements on workpieces as described above. Measured geometric deviations have remained within the tolerances of class H in table 16.4 (Figure 16.9).

Measured deviations have not shown any relationship to

• diameter or length (assessed up to diameter 900 mm);

---

*Circular run-out tolerances according to ISO 1101 do not limit the deviations of the straightness of cylinder generator lines or the flatness of faces, which is in contrast to total run-out tolerances.

Table 16.4:   General tolerances on circular run-out (values in mm)

Tolerance class	Run-out tolerance
H	0.1
K	0,2
L	0.5

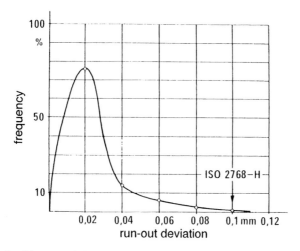

Figure 16.9:   Measured circular radial run-out deviations.

- size tolerance;

- specified surface roughness;

- material (assessed metallic material).

This has been proved during use since 1974 in, for example the German industry.

The reason for this is probably the following. The general tolerances were set up depending on the greatest measured deviations. These deviations were not due to the inaccuracy of the machine tool but rather to inaccuracy when adjusting the workpiece in the machine tool after the workpiece has been turned over (rechucking). During adjustment, the workpiece was usually measured where the greatest deviations are to be detected (e.g. at the outer diameter). Therefore small and large workpieces were adjusted with the same inaccuracy.

16.3.11   Total run-out

The radial total run-out deviation consists of three components: circular run-out deviation, straightness deviation and parallelism deviation. The axial total run-out deviation consists of two components: circular run-out deviation and flatness deviation. Each of these is controlled by its general tolerance.

General tolerances on radial total run-out are not intended to be standardized, for similar reasons as for cylindricity. General tolerances on axial total run-out are not intended to be standardized, since general tolerances on perpendicularity are already standardized.

16.4   DATUMS

For general tolerances of orientation, location and run-out it is necessary to determine the datums without drawing indications. According to ISO 2768-2, the longer of the two considered features applies as the datum. When they are of equal length, either may serve as a datum.

An exception applies with general tolerances of run-out when there are bearing surfaces designated as such in the drawing. Then these surfaces serve as the datum(s).

Although, with the exception of designated bearing surfaces, datums for general geometrical tolerances are not designated in the drawing, there is no accumulation of general geometrical tolerances possible from one feature to the next etc., because the general geometrical tolerances apply to all possible combinations of features of the workpiece.

16.5   INDICATION ON DRAWINGS

When the general tolerances according to ISO 2768-2 should apply, this has to be designated in or near the title block, for example

General Tolerances   ISO 2768-mH

In this example m stands for general dimensional tolerances class m and H for general geometrical tolerances class H.

The indication on drawings when the envelope requirement (Rule #1 of ANSI Y14.5M or DIN 7167) is used in addition to the general tolerances as a general requirement is dealt with in 16.6.

16.6   ENVELOPE REQUIREMENT IN ADDITION TO GENERAL FORM TOLERANCES

It was a strong desire from some countries, at least for a transition period, to allow in addition to the general tolerances the use of the envelope requirement (Rule #1 of ANSI Y14.5M or DIN 7167) without individual indications for all single features of size. (Single features of size consist of a cylindrical surface or two parallel opposite plane surfaces.) This means limiting form deviations also by the envelope requirement even when the feature does not have the function of a fit and the feature's function would allow certain form deviations when it is everywhere at maximum material size.

In these cases the drawing should be designated in or near the title block, for example

General Tolerances   ISO 2768-mH-E

With this designation, the envelope requirement applies to all individual features of size, provided that there is no individually indicated straightness tolerance applied that is greater than the size tolerance. The envelope requirement is considered as an additional requirement. All the other indications (dimensional and geometrical tolerances) retain their meaning as described above (Figure 16.10). See also 17.3.

Drawing indication: Interpretation:

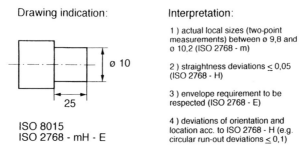

1) actual local sizes (two-point measurements) between ø 9,8 and ø 10,2 (ISO 2768 - m)

2) straightness deviations $\leq 0{,}05$ (ISO 2768 - H)

3) envelope requirement to be respected (ISO 2768 - E)

4) deviations of orientation and location acc. to ISO 2768 - H (e.g. circular run-out deviations $\leq 0{,}1$)

ISO 8015
ISO 2768 - mH - E

Figure 16.10: Envelope requirement as general requirement, example according to ISO 8015 and ISO 2768-2.

16.7   APPLICATION OF GENERAL GEOMETRICAL TOLERANCES ACCORDING TO ISO 2768-2

When the general geometrical tolerances according to ISO 2769-2 should apply, this has to be stated on the drawing (see 16.5).

The general tolerances according to ISO 2768-2 are applicable both when the principle of independency applies and when the principle of dependency applies (see 16.6 and 17).

When the principle of dependency should apply, see 16.6.

For the application of general geometrical tolerances according to ISO 2769-2 the normal workshop accuracy should be known. The general geometrical tolerances should be not smaller than the normal workshop accuracy. The latter can be assessed by measurements of workpieces manufactured under normal workshop (production) conditions. The normal workshop accuracy (regarding geometrical deviations) depends on the accuracy of the (most inaccurate) machine tools and on the accuracies of adjustment of the workpieces during re-chucking on the machine tool.

Normally the normal workshop accuracy in mechanical engineering corresponds to ISO 2768-H (and to ISO 2758-mH when general tolerances for dimensions are included).

Table 16.5: General geometrical tolerances for machining according to ISO 2768-2 synoptic table

Characteristic	Tolerance
—	Table 16.1
▢	Table 16.1
○	Size tolerance or Table 16.4*
//	Size tolerance or Table 16.1†
⊥	Table 16.2
⹀	Table 16.3
∕	Table 16.4
◎	See ∕
⌀	See ○ − //
⌒⌀	Radial, see ∕ − //
	Axial, see ⊥
⌒ ⌓ ∠ ⊕	Not specified‡

*Whichever is the smaller.
†Whichever is the greater.
‡Needs individual indication.

16.8 GENERAL TOLERANCES FOR CASTINGS

International standards on general geometrical tolerances for castings have not yet been published, but are in preparation. Measurements on castings in the industries of various countries have been made and submitted to the committee ISO/TC 3 in order to derive an appropriate tolerance system for general tolerances. This system is planned under the principle of independency ISO 8015, because,

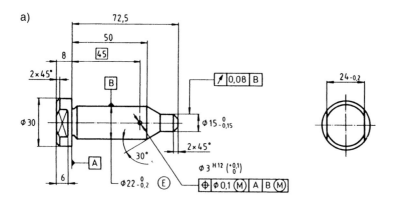

GENERAL TOLERANCES ISO 2768-MH

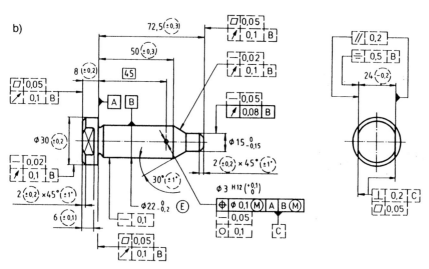

Figure 16.11: Example of applying general tolerances: (a) drawing indication; (b) interpretation. (Tolerances shown in dashed lines (boxes and circles) are general tolerances according to ISO 2768-mH, which will be automatically respected by machining in a workshop with a customary accuracy equal to or finer than ISO 2768-mH and would not normally need to be inspected.)

Table 16.6: General tolerances of straightness, flatness and parallelism for welded parts according to DIN 8570

Tolerance class	Range of dimension (greater length of the surface) (in mm)									
	≤ 30, ≤ 120	> 120, ≤ 400,	> 400, ≤ 1000	> 1000, ≤ 2000	> 2000, ≤ 4000	> 4000, ≤ 8000	> 8000, ≤ 12 000	> 12 000, ≤ 16 000	> 16 000, ≤ 20 000	> 20 000
	Tolerances (in mm)									
E	0.5	1	1.5	2	3	4	5	6	7	8
F	1	1.5	3	4.5	6	8	10	12	14	16
G	1.5	3	5.5	9	11	16	20	22	25	25
H	2.5	5	9	14	18	26	32	36	40	40

for example, flatness deviations and size deviations have different causes and are therefore independent of each other.

16.9 GENERAL TOLERANCES FOR WELDED PARTS

An International standard on general geometrical tolerances for welded parts is in preparation. Probably it will be similar to the German Standard DIN 8570 Part 3 on general tolerances of straightness, flatness and parallelism for welded parts. Table 16.6 gives these tolerances. They apply under the principle of independency ISO 8015.

17 Tolerancing Principles

17.1 LIMITATION BY FUNCTION

The function of a feature limits its permissible deviations of size, form, orientation and location. If one of these deviations is too large, it affects the functioning of the feature. This applies to each feature, and may be illustrated by the following example of the flange shown in Figure 17.1.

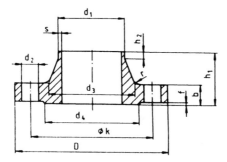

Figure 17.1: Flange: D, reduction limited by distance to the hole d_2, increase limited by adjoining parts; d_2, reduction limited by bolt, increase limited by pressure under the nut; r, reduction limited by fatigue notch effect, increase limited by nut face; d_3, reduction limited by strength, increase limited by nut face; and so on.

The deviations allowed by function may sometimes be relatively great (e.g. for D in Figure 17.1), but limits always exist. Otherwise the considered feature would not need to exist.

17.2 NEED FOR COMPLETELY TOLERANCED SHAPE

Consequently the tolerance for each characteristic should be given. As there is hardly any communication other than by technical drawings, reliance on undefined workshop accuracy or good workmanship is not sufficient. The international exchange of technical drawings requires that the drawing specify all dimensional and geometrical tolerances necessary to define the shape and size of the part completely.

179

17.3 SITUATION IN THE PAST

Regarding geometrical tolerances, this requirement has not always been respected. In the past, in most countries there has been reliance on undefined good workmanship regarding deviations of perpendicularity, coaxiality and symmetry for features that have no corresponding tolerance indication on the drawing.

In some countries (but not others) the form deviations have been assumed to be controlled by the dimensional (size) tolerances. This is specified for example by Rule #1 of the US Standard ANSI Y14.5M. This requires that the feature (cylindrical surface or two parallel oppsite plane surfaces) shall be contained within an envelope of perfect form at its maximum material size (maximum size of a shaft or minimum size of a hole). In other words, a shaft at maximum material size everywhere should have neither straightness nor circularity deviations, i.e. it should be of perfect form.

This restriction is necessary for fits, but in most other functional cases it is unnecessary. Reviews of technical drawings in different countries have revealed that generally less than 10% of features (dimensions) on a drawing should be associated with Rule #1. More than 90% of features do not need this restriction. Their function allows geometrical deviations even when the feature is at its maximum material size everywhere.

For features for which Rule #1 is not required a disclaimer should be indicated on the drawing. However, this has usually not been done, because it would have embroidered the drawing with disclaimers. On the other hand the drawing should point out what is required, rather what is not. (When disclaimers are indicated and a disclaimer is missing it might be doubtful whether Rule #1 is not in fact required or the draughtsman has simply overlooked this dimension among the many disclaimers to be indicated in the drawing.)

In the inspection of workpieces the requirement of Rule #1 has been checked only if a full form gauge was used. However, in most cases snap gauges or measuring instruments have been used, and the requirements of Rule #1 have not been verified (Figure 17.2).

In order to respect Rule #1, the feature must be manufactured with actual local sizes that depart from the maximum material size by the amount of the expected straightness and certain circularity deviations. In other words, one is not allowed to take full advantage of the dimensional tolerance for the actual (local) sizes. Frequently this has not been respected.

From measured workpieces it is known that the magnitude of form deviations permitted by Rule #1 is not in line with the normal workshop accuracy of various industrial companies. The form deviations that occur through normal machining processes are sometimes greater than small dimensional tolerances. 9500 measurements obtained in 19 companies in Switzerland revealed that in 19% the straightness deviation was greater than the dimensional tolerance. As there are also roundness deviations and variations in size to be taken into account,

Drawing indication:

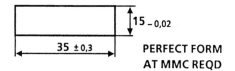

15 _{-0,02}

35 ±0,3   PERFECT FORM
AT MMC REQD

Interpretation:

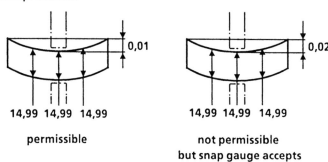

0,01   0,02

14,99 14,99 14,99   14,99 14,99 14,99

permissible   not permissible
                    but snap gauge accepts

Figure 17.2: Limitation of straightness deviations by size tolerance.

much more than 19% violate Rule #1, although the companies pretended to manufacture according to it. From measurements in Germany, Japan and the UK the same is found. On the other hand, some dimensional tolerances (e.g. h14) are much greater than the form deviations that occur through normal machining processes (see Table 16.1). Therefore general form tolerances, as described below, would be more in line with the normal manufacturing capability than reliance on Rule #1.

This has led to the relationship between dimensional and geometrical tolerances being described and defined in different ways in various countries. Some have adopted Rule #1; others have treated dimensional and geometrical tolerances independently of each other. Some countries have regarded an indicated geometrical tolerance to a feature as a disclaimer of Rule #1; other countries have not. In some countries it has been assumed that the feature should not violate the boundary of perfect form at least material size, while in others it has not. Some countries have regarded a dimensional tolerance adjacent to a positional toleranced pattern as different from a dimensional tolerance adjacent to a dimensional tolerance, while others have not.

Even within companies, opinions have varied. Generally while designers have tended towards the thinking of Rule #1, inspectors have referred to the fact that usually size and form (and orientation and location) are inspected independently of each other.

17.4 PRINCIPLE OF INDEPENDENCY

As this situation was considered unsatisfactory, the International Standardizing Committees ISO/TC10/SC5 "Technical Drawings, Dimensioning and Tolerancing" and ISO/TC3 "Limits and Fits" decided to standardize in ISO 8015 the principle of independency as the fundamental tolerancing principle. It states that

"Each requirement for dimensional or geometrical tolerancing specified on a drawing shall be met independently, unless a particular relationship is specified, i.e. Ⓜ or Ⓔ or Ⓛ.

This means that linear dimensional tolerances (e.g. ± 0.1 or H7) control only actual local sizes (two-point measurements) (e.g. wall thicknesses), but not the form deviations of the feature (Figure 17.3).

Drawing indication: Interpretation:

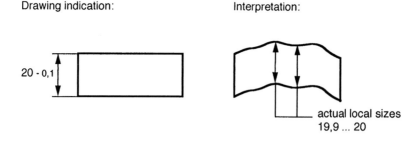

$20 - 0,1$

actual local sizes
19,9 ... 20

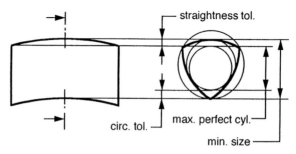

straightness tol.

circ. tol.

max. perfect cyl.

min. size

Figure 17.3: Principle of independency; linear dimensional (size) tolerance.

Angular dimensional tolerances, specified in angular units (e.g. ± 1°), control the general orientation of lines or line elements of surfaces, but not their form deviations. The general orientation of the line derived from the actual surface is the orientation of the contacting line of perfect form (contacting straight line). The maximum distance beween the contacting line and the actual line should be the least possible (Figure 3.42).

When no relationship is specified the geometrical tolerance applies regardless of feature size, and the two characteristics are treated (inspected) as unrelated requirements (Figures 3.42, 17.3–17.5).

Drawing indication:

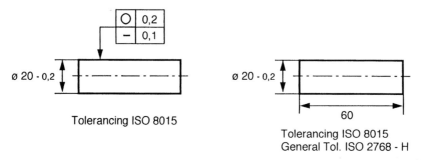

Tolerancing ISO 8015

Tolerancing ISO 8015
General Tol. ISO 2768 - H

Interpretation, extreme permissible part:

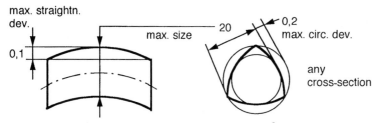

Figure 17.4: Principle of independency: size and form.

Consequently the principle of independency demands a separate indication for each requirement (which may cause a particular checking operation and may be inspected regardless of other characteristics). Therefore **it is sufficient to inspect what is indicated in the drawing**. The requirements are indicated in an analytical way. They may be

● linear dimensional tolerance (ISO 8015, ISO 286);

Tolerancing Principles

Drawing indication:

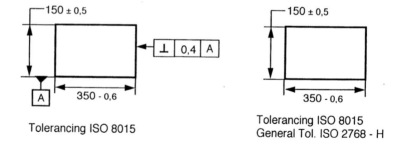

Tolerancing ISO 8015

Tolerancing ISO 8015
General Tol. ISO 2768 - H

Interpretation, extreme permissible part:

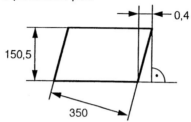

Figure 17.5: Principle of independency; size and orientation.

- angular dimensional tolerance (ISO 8015);
- form tolerance (ISO 1101);
- tolerance of orientation, location and run-out (ISO 1101);
- envelope requirement Ⓔ (ISO 8015, ISO 286);
- maximum material requirement Ⓜ (ISO 2692);
- least material requirement Ⓛ (ISO 2692);
- reciprocity requirement Ⓡ (ISO 2692);
- projected tolerance zone Ⓟ (ISO 10 578);
- tolerances in the restrained condition ISO 10 579, tolerances in the free state Ⓕ.

There are no "hidden rules" (like e.g. Rule # 1), i.e. there are no rules (requirements) that are not explicitly indicated on the drawing.

As, according to the principle of independency form deviations are not controlled

184

by the size tolerances, it calls for the application of general tolerances of form. However, the application of general geometrical tolerances is necessary anyway in order to control deviations of orientation and location (see 16).

Drawings to which general geometrical tolerances are not applied and that do not specify all necessary geometrical tolerances are incomplete. They may cause difficulties when the designer has assumed a more precise normal workshop accuracy than the manufacturer has implied.

The application of the principle of independency (ISO 8015) together with general geometrical tolerances (e.g. ISO 2768) and the symbol Ⓔ, where applicable, provides the following advantages:

- the same interpretation of drawings worldwide (no misunderstandings due to different national rules;

- drawings can be toleranced according to the functional requirements (no unnecessary limitations like e.g. Rule #1);

- drawing indications show what is required, and not what is not required (no disclaimer necessary);

- drawings are decisive (fully toleranced);

- tolerancing in accordance with inspection practices (analytical method, each check requires a separate requirement indication, no hidden rules);

- drawings are clearer (they are not embroidered with tolerance indications);

- there is a foundation for economical production.

General geometrical tolerances together with the symbols Ⓜ, Ⓛ, Ⓡ, Ⓟ and Ⓕ, where applicable, should be applied in any case (also when the principle of dependency is applied; see 17.6).

It should be noted that in cases of extreme lobed form deviations the feature cross-section can violate the circle of maximum material size (Figure 17.6). The extreme permissible part with maximum material size is shown in Figure 17.4. When this is not allowed (e.g. in the case of a fit), the envelope requirement should be applied (see 10).

The maximum perfect cylinder that is contained in the smallest permissible cylindrical part (shaft) is smaller than the least material size by the sum of the tolerance on circularity and the tolerance on straightness because of the lobed and bent form deviation (Figure 17.7). This has to be taken into account, for example, with stock material.

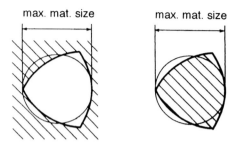

Figure 17.6: Violation of the maximum material size circle by a lobed form.

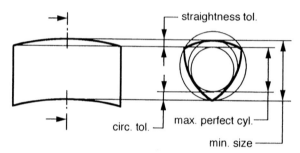

Figure 17.7: Maximum perfect cylinder (shaft) contained in the least part.

17.5 IDENTIFICATION OF DRAWINGS

Drawings to which the principle of independency applies must be distinguished from those to which other (e.g. former) rules apply, and are therefore to be identified according to ISO 8015 by the title box indication:

Tolerancing ISO 8015.

17.6 PRINCIPLE OF DEPENDENCY

Some national standards provide a tolerancing concept, the principle of dependency, as an alternative concept to the principle of independency according to ISO 8015.

The principle opposite of dependency specifies for features of size (i.e. established by cylindrical surface or two opposite parallel plane surfaces) the envelope requirement without any drawing indication (e.g. ANSI Y14.5 M, Rule #1, BS 308, DIN 7167; see 21).

This applies to size and form of isolated features only. The principle of dependency does not apply to orientation or location of related features. For example, a cube consists of three individual features of size, each composed of a set of two opposite parallel plane surfaces. The perpendicularity of those individual features of size is not controlled by their size tolerances. It is controlled by general geometrical tolerances, if applied, or by individually indicated perpendicularity tolerances (see e.g. Figure 20.63).

The principle of dependency does not apply to angular dimensions or angular dimensional tolerances (e.g. $\pm 1°$). Most national standards (e.g. BS 308) define angular dimensional tolerances as limits for contacting lines, as defined in ISO 8015 (Figure 3.42), i.e. straightness and flatness deviations are not controlled by the angular tolerance. However, ANSY Y14.5M defines angular tolerances differently (see 21.1.2).

The principle of dependency can be indicated on drawings, when general geometrical tolerances according to ISO 2768 are applied, by the letter "E" in the title box identification (e.g. ISO 2768-mH-E). See 16.5.

17.7 CHOICE OF TOLERANCING PRINCIPLE

For new or altered drawings it must be decided whether the principle of independency (ISO 8015) or the principle of dependence (ISO 2768 - ... - E, or national standard, e.g. ANSI Y14.5M Rule #1, BS 308, DIN 7167) should be applied. See also 21.

The principle of dependency has the advantage that the designer does not need to check where the envelope requirement is necessary. In this respect the principle of dependency is failsafe.

In addition, with the principle of dependency the following must be observed:

- To respect the envelope requirement, the manufacturer must decrease the tolerance at the maximum material side by the amount of the expected form deviations. According to the drawing, this applies to all features of size, even when there is no functional necessity.

- With the inspection, it is not sufficient to inspect the dimensional deviations and the form deviations. It must also be checked whether the envelope requirement has been respected (see 10 and 18.7.11).

- The principle of dependency tempts the designer to specify narrow dimensional tolerances, although wider dimensional tolerances and (narrower) geometrical tolerances would be sufficient.

- Because it is well known that the envelope requirement with the principle of dependency is often violated, there is a danger that the designer might specify narrower dimensional tolerances than are necessary according to the function.

These all lead to unnecessarily great effort (and cost) in manufacturing and

187

inspection, or to a large number of workpieces that do not comply with the drawing.

In contrast, the principle of independency ISO 8015 requires slightly more effort in design, but allows function-related designs with all advantages in manufacturing and inspection. The envelope requirement must be respected only where it is a functional need. This is a prerequisite for economical production taking into account the requirements of quality assurance according to ISO 9000 etc. Figure 17.8 gives a comparison of the two principles.

	Principle of	
	independency ISO 8015	dependency
Design	Ⓔ to be indicated (thinking necessary)	Failsafe (no thinking necessary)
	Decreasing of tolerance	
Manufacturing	With Ⓔ only	Always*
	Check of envelope requirement	
Inspection	With Ⓔ only	Always*

*or unnecessary narrow dimensional tolerance.

Figure 17.8: Comparison of the principle of independency ISO 8015 with the principle of dependency.

Regarding the application of general geometrical tolerances (e.g. according to ISO 2768) and of Ⓜ, Ⓛ, Ⓡ, Ⓟ and Ⓕ, there is no difference between the two principles.

18 Inspection of Geometrical Deviations

18.1 GENERAL

The following deals with generalities on the inspection of geometrical deviations. A synopsis of inspection methods is given in ISO TR 5460. More detailed descriptions are given in the former East German Standards TGL 39 092 to TGL 39 098 and TGL 43041 to 43045, TGL 43 529 and TGL 43 530.

Geometrical tolerances are geometrically exactly defined. They determine geometrical zones within which the surface of the feature must be contained.

There are several methods for inspecting whether the geometrical tolerances have been respected. These methods are more or less precise. The drawing indications according to ISO 1101 do not prescribe a particular (certain) inspection method. According to the German Standard DIN 2257 T.2, the inspection method should be selected so that the measuring uncertainty u_{95} is between 0.1 and 0.2 times the manufacturing tolerance ($\pm u_{95}$ = range within which the unknown true value is contained with a statistical probability $P = 95\%$). See 18.11.

In industrial practice the inspection of geometrical deviations is often only economical with less precise inspection methods. Less precise inspection methods are normally more time-consuming and less costly. In contrast, more precise inspection methods are normally more time-consuming and less costly. Therefore it is often advisable to start the inspection with a cheap but less precise method and to switch to a more precise (and more expensive) method only in cases where the measurement result is near the limit given by the tolerance (see e.g. 18.7.9.3.6).

A prerequisite is that the possible errors in the inspection methods are known. Information on systematic errors of inspection methods are given in the East German Standards mentioned above.

Specifications on the necessary sample sizes (number of traces and number of probed points) are not internationally standardized. They depend on the size of the feature to be inspected and on the ratio between form deviation and geometrical tolerance. The measurement result shall be representative of the feature. Some hints are given in 18.9.

Particular inspection methods should not be prescribed by the drawing indication for the following reasons.

- The type and frequency of inspection to be used depend on the control of the manufacturing process.

- There are often different but equivalent correct inspection methods. Prescription of particular inspection methods would force the manufacturer to provide inspection devices prescribed by the customer, although other sufficient inspection devices are already available.

- There may be a contradiction between the drawing indication according to ISO 1101 (geometrically defined tolerance zone) and the effect assessed by the inspection method.

- Prescribing inspection methods that differ in assessment from the precise tolerance zone requires further specifications of the measurement conditions. Inspection methods that differ from the precise tolerance zone and different measurement conditions would make the inspection of geometrical deviations difficult and prone to mistakes.

Therefore the ISO Standards do not recommend the indication of a particular inspection method. The drawing should only specify the geometrically exactly defined requirement (tolerance zone) (see 19).

18.2 TERMS

Embodiment Measuring is comparing. In order to measure geometrical deviations, the workpiece surface must be compared with a geometrical ideal feature. But the latter features cannot be manufactured. Therefore almost geometrical ideal embodiments of geometrical ideal features (e.g. straight lines, planes, circles, cylinders and spheres) are used. They are called embodiments in the following.

The embodiments can be surfaces of measuring devices (e.g. straight edges, measuring tables, solid angle and sliding guides of measuring devices) or can be established by the movements of precision guides (e.g. by rotating the workpiece or the measuring device).

Datum This is a theoretical exact geometrical reference (such as an axis, plane or straight line) to which toleranced features are related. Datums may be based on one or more datum features of a workpiece; ISO 5459 (see 3.4).

Datum system This is a group of two or more separate datums used as a combined reference for a toleranced feature; ISO 5459 (see 3.4).

Datum feature This is a real feature of a workpiece (such as an edge, a plane surface or a hole) that is used to establish the location of a datum; ISO 5459 (see 3.4).

Datum target This is a point, line or limited area on the workpiece to be used for contact with the manufacturing and inspection equipment, to define the required datums in order to satisfy the functional requirements; ISO 5459 (see 3.4).

Simulated datum feature This is a real surface of sufficiently precise form (such as a surface plate, a bearing or a mandrel) contacting the datum feature(s) and used to establish the datum(s). Simulated datum features are used as the **practical embodiment** of the datums during manufacture and inspection; ISO 5459 (see 3.4).

Reference element This is a geometrical ideal element (a straight line, circle, line of any form defined by theoretical exact dimensions, plane, cylinder, or a surface of any form defined by theoretical exact dimensions) relative to which the geometrical deviations are evaluated.

The reference element is established by or derived from the embodiment at the measured (toleranced) feature. When the reference element is calculated (e.g. using a form measuring instrument), the reference element may differ from the embodiment in orientation, location and size.

The orientation and location of the reference element depend on the characteristic to be measured (deviations of unrelated form, related form, orientation, circular or total axial run-out, location, circular or total radial run-out); see below, geometrical deviation.

Coordinate measuring machines and form measuring instruments often approximate the reference element by the (Gaussian) least-squares substitute element (see 8.1).

The substitute element intersects or contacts the workpiece surface, but the reference element need not (see Figure 8.5 and Table 3.3). In cases of a straight line or plane the minimum zone substitute element and the reference element have the same orientation but not necessarily the same location. In cases of a circle or cylinder the minimum zone substitute element and the reference element have the same location and orientation but not necessarily the same size.

Geometrical deviation The definitions of geometrical deviations are not yet internationally standardized. But they can be derived from the definitions of geometrical tolerances according to ISO 1101.

The geometrical deviation is the deviation of a workpiece feature (axis, section line, edge, surface or median face) from the (geometrical ideal) reference element (embodiment). The reference element is

- orientated according to the minimum requirement (18.3.2) for the measurement of deviations of unrelated form;

- orientated according to the datum (18.3.4) or datum system (3.4) for the measurement of deviations of related form, orientation, circular or total axial run-out;

- orientated and located according to the datum (18.3.4) or datum system (3.4) and to the theoretical exact dimensions for the measurement of deviations of related form, location, circular or total radial run-out.*

For the evaluation of the geometrical deviation to be compared with the geometrical tolerance see 18.6.

18.3   ALIGNMENT OF THE WORKPIECE

18.3.1   General

Most important for measurement is the alignment of the workpiece in the measurement device. Misalignment can cause large errors in measurement, so that workpieces that actually comply with the specification may appear not to, or more seldom but in principal possibly, workpieces that actually do not comply with the specification may appear to do.

For the measurement of form deviations the minimum requirement has to be respected (see 18.3.2 and 18.3.3).

For the measurement of deviations of orientation, location and run-out (related geometrical deviations) the minimum-rock requirement must be respected for the datum(s) (see 18.3.4).

For the measurement of roundness or cylindricity additional alignment requirements must be respected (see 18.3.4).

18.3.2   Minimum requirement

The minimum requirement defines the orientation or location of the form tolerance zone. With straightness and flatness, it requires that the parallel straight lines or parallel planes enclosing the feature be directed so that their distance is a minimum (Figure 18.1); ISO 1101.

With roundness and cylindricity, it requires that the concentric circles and the coaxial cylinders enclosing the feature be located so that their radial distance is a minimum (Figure 18.2); ISO 1101.

Deviations from the minimum requirement (e.g. alignment according to the Gaussian regression line) simulate larger form deviations.

---

*In cases of circular radial run-out or total radial run-out the theoretical exact dimension between the axes of the toleranced feature and the datum feature is zero.

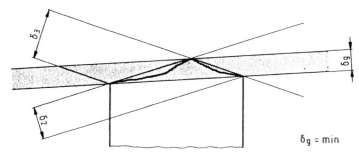

Figure 18.1: Minimum requirement for straight lines.

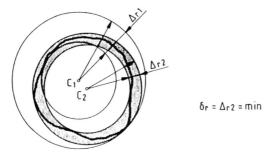

Figure 18.2: Minimum requirement for circles.

18.3.3 Additional alignment requirements for the measurement of roundness or cylindricity deviations

18.3.3.1 Axis of measurement

The **axis of measurement (reference axis)** is a straight line relative to which the measurement is performed and established by the measuring device. The measurements are considered to be perpendicular and centred on the axis of measurement.

The (desired) **workpiece established axis** is a straight line about which the relevant part of the workpiece is required to surround and may be defined (established) as follows.

(a) A straight line such that the root mean square value of the distances from it to the defined centres of a representative number of cross sections has a minimum value. The defined centres may be the centres of the

● regression circle (by Gauss, the circle of which the sum of the squares of the radial distances to the circumference is a minimum, namely the least-squares circle, or LSC);

- minimum-zone circle (by Chebychev, the circle of which the maximum radial distance to the circumference is a minimum, namely the MZC);

- contacting circle (the maximum inscribed circle for holes, MIC, the minimum circumscribed circle for shafts, MCC, sometimes refered to as the plug gauge circle, PGC, and ring gauge circle, RGC).

(b) A straight line passing through the defined centres of two separated and defined cross-sections. For the defined centres see (a).

(c) A straight line passing through the defined centre of one defined cross-section and perpendicular to a defined shoulder. For the defined centre see (a). The defined shoulder may be determined by the

- regression plane (by Gauss, the plane of which the sum of the squares of the distances to the actual feature (shoulder) is a minimum);

- minimum-zone plane (by Chebychev, the plane of which the maximum distance to the actual feature (shoulder) is a minimum);

- datum plane (according to the minimum-rock requirement, see 18.3.4).

(d) A straight line passing through two support centres (this axis is independent of the surface of the workpiece).

(e) An axis of two coaxial cylinders with minimum separation from the surface irregularities of the workpiece.

(f) A substitute cylinder axis (see 8.1).

If not otherwise specified (agreed upon) according to ISO 1101, the definition (e) applies. However, current practice is, for practical reasons, to use one of the other possibilities, preferably one of the Gaussian methods.

18.3.3.2 Inclination of workpiece axis

If the (desired) workpiece established axis is inclined to the axis of measurement (reference axis), a round workpiece feature (about the desired workpiece established axis) may appear oval and an oval workpiece feature may appear round (Figure 18.3).

The planned ISO Standards on measurement of roundness and cylindricity will probably specify that the levelling (alignment) should be performed at the top and bottom of the workpiece feature until the least-squares centres (LSC) at these

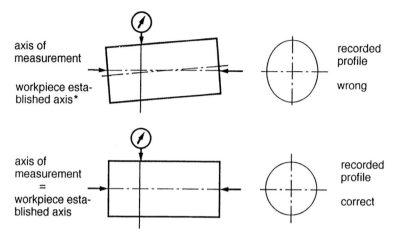

* desired but not achieved in the measurement

Figure 18.3: Roundness measurement: false measurement caused by a plane of measurement not perpendicular to the workpiece established axis (workpiece established axis inclined to the axis of measurement).

two levels coincide to better than 10% of the roundness or cylindricity tolerance. For practical reasons with small roundness or cylindricity tolerances this value may be greater.

18.3.3.3 Offset of workpiece axis

Form measuring instruments highly magnify the form deviations but not the radius itself. When these instruments are used and when if in the measuring plane the centre of the workpiece cross-section line (profile) differs from the axis of measurement (reference axis), distortions of the profile occur (limaçon effect) (Figure 18.4).

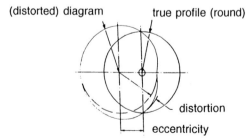

Figure 18.4: Distortion of the profile diagram caused by an offset of the true workpiece centre from the axis of measurement (limaçon effect).

The planned ISO Standard on the measurement of roundness will probably specify that the workpiece should be centred at each level that is measured individually to improve the resolution of the measurement and minimize the limaçon effect. The centring should be continued until a measuring range can be used (i.e. the profile is fully contained in this range) that results in a resolution of the instrument that is smaller than 1% of the roundness tolerance.

18.3.4   Minimum-rock requirement

In the measurement of deviations of orientation, location and run-out, the minimum-rock requirement applies to the alignment of the datum feature.

When the datum feature is not stable (convex datum feature) relative to the contacting surface (simulated datum feature, e.g. measuring table or mandrel), it should be arranged so that the possible movement (inclination) in any direction is equalized, i.e. so that the maximum possible inclination to the extreme position is a minimum (minimum-rock requirement) (Figure 18.5). In other words, the datum feature should be aligned relative to the simulated datum feature into a median position.

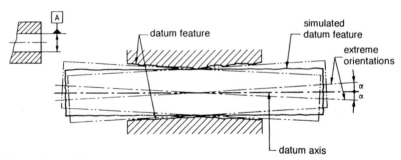

Figure 18.5:   Minimum-rock requirement for cylindrical datum features.

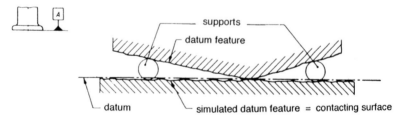

Figure 18.6:   Minimum-rock requirement for plane datum features.

In order to arrange this in the case of a plane datum feature, three equal-height supports should be placed between the datum feature and a simulated datum feature at the same distance from the end of the datum feature (Figure 18.6).

In the case of a cylindrical datum feature the contacting cylinder (the maximum inscribed simulated datum cylinder for a datum feature that is a hole, and minimum circumscribed simulated datum cylinder for a datum feature that is a cylindrical shaft) should be arranged according to the minimum-rock requirement, i.e. in a median position. The same applies to a datum feature that is an axis.

When the datum feature is the common axis of two cylindrical shafts of different nominal sizes according to ISO 5459, the datum is the common axis established by the two smallest circumscribed coaxial cylinders (Figure 18.7). This definition is not precise. Figure 18.7 is taken from ISO 5459, and shows the case in which the **sum** of the diameters of the coaxial simulated datum cylinders is a minimum.

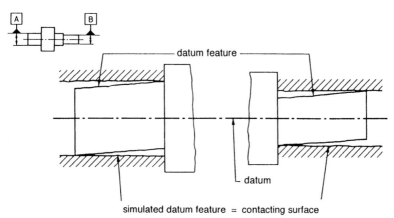

Figure 18.7: Common axis as a datum (ISO 5459).

Often the connection of the centres (centres of the regression circles) of the cross-sections at half the length of the datum features should establish the common axis (Figure 18.8) (measurement with form measuring instrument or a support in edge V-blocks). See also 18.7.10.1.

In order to avoid disputes, when small geometrical tolerances and common axis are required, the method for establishing the common axis should be specified in the drawing, e.g. the support at specified cross-sections (Figure 18.8).

For the alignment of the datum features according to the three-plane concept see 3.4.

The minimum-rock requirement may cause difficulties when the actual surface of a plane datum surface is convex and, for example, more inclined near one of the edges. Then the extreme rock positions and the minimum-rock position may be unreasonable. Therefore discussions are presently under way in the ISO and

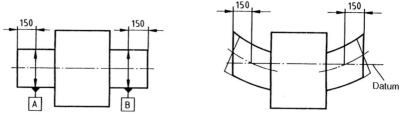

Figure 18.8: Common datum axis established by the centres of two specified cross-sections.

the ANSI on whether the minimum-rock requirement should be replaced by, for example, the least-sqares method or a candidate datum method. According to the least-squares method, the substitute element gives the orientation of the datum. According to the candidate datum method, contacting planes that contact the actual datum feature to a certain extent give possible orientations for which, if possible, all related tolerances are respected. ISO 5459 will be amended accordingly.

18.4 INTERCHANGING DATUM FEATURE AND TOLERANCED FEATURE

With related geometrical tolerances, the drawing distinguishes between toleranced and datum features. When, for inspection, the toleranced and datum features are interchanged, completely different values of the geometrical deviation may be obtained from the same workpiece (Figure 18.9).

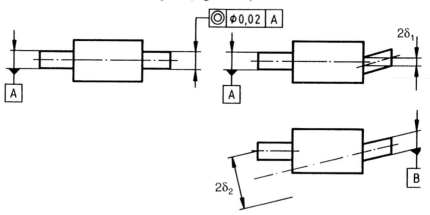

Figure 18.9: Change in geometrical deviation caused by interchanging of toleranced and datum features.

Therefore, for inspection, toleranced and datum features must not be inter-

198

changed. When the datum feature indicated on the drawing is not suitable for inspection purposes (e.g. because it is too short), a change in the drawing is necessary (Figure 19.13).

18.5 SIMPLIFIED INSPECTION METHOD

Inspection of certain types of geometrical tolerances (e.g. the coaxiality tolerance) is relatively costly. Often in such cases in the first step a "quick" (but less precise) inspection is chosen and only in case of doubt is the more precise (and more costly) inspection executed in a second step.

For example, in the cases of a coaxiality tolerance or a common straightness tolerance zone of axes, the inspection in the first step is performed as if there were a run-out tolerance of the same value. Only when this tolerance is exceeded is it checked whether the coaxiality tolerance or the straightness tolerance of the axis is exceeded.

Often with related geometrical tolerances, precise verification of the minimum-rock requirement is very costly.Therefore approximate inspection methods are used with the aid of V-blocks, mandrels, centre holes etc. Using V-blocks, the form deviation of the datum feature leads to simulation of a larger related geometrical deviation than actually exists (according to the definition). Depending on the shape of the form deviation and the angle of the V-block, the increase can be equal to or less than the form deviation of the datum feature.

A similar consideration applies with the use of centre holes. There the eccentricity of the centre hole relative to the datum feature increases the result of the measurement of the related geometrical deviation.

Mandrels for inspection purposes are normally rated in diameter in units of 0.01 mm. The largest mandrel that fits into the hole should be used. In cylindrical holes the mandrel can be inclined by 0.01 mm at most (Figure 18.10), and thereby give and incorrect measurement result. In very rare cases of very detrimental form deviations larger inclinations may occur (Figure 18.11).

When the indicated excess of the tolerance is of the same magnitude as the possible error caused by the inspection equipment and significant compared with the tolerance, more precise inspections should be executed (or the inspection method should be agreed upon between parties).

Pneumatic mandrels are self-centring, as are expanding mandrels and conical mandrels.

18.6 EVALUATION OF MEASUREMENT

Evaluation of a measurement can be manually by calculation, or graphically or automatically by suitable computer programs (e.g. in a form measuring instrument or a coordinate measuring machine). In the following sections the principles of evaluation by calculation are described.

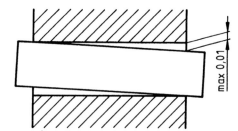

Figure 18.10: Mandrel: possible inclination.

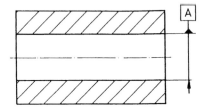

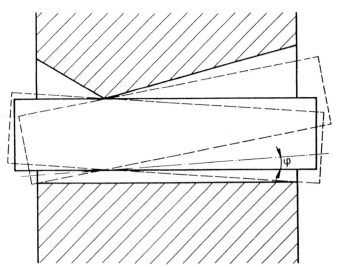

Figure 18.11: Mandrel: possible deviation form the minimum-rock requirement.

In many cases a cylindrical tolerance zone is derived from the function, and is indicated in the drawing. When during inspection the deviations in cartesian (rectangular) coordinates are assessed, they can be compared with the tolerance diameter with the aid of the diagram (Figure 18.12).

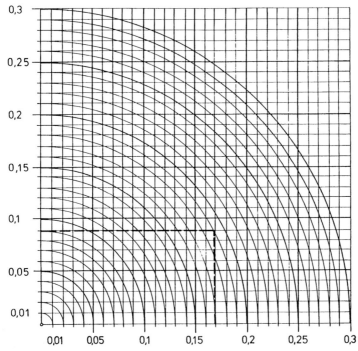

Figure 18.12:   Diagram to compare assessed coordinates with cylindrical or circular tolerances. For example, measured coordinates 0.17 and 0.09 lie within Ø 0.4 ($r = 0.2$) but outside Ø 0.36.

Figure 18.13 shows a typical method for the assessment of related geometrical deviations (here parallelism deviations of an axis relative to a datum axis) by detecting rectangular coordinates δ_x, δ_y and calculation of the cylindrical coordinate δ_p.

Often the tolerance applies over the length l of the feature, whereas for practical reasons the measurement was applied to the length l_m. The measured values are to be corrected by the ratio l/l_m, i.e.

$$\delta_u = \delta_m \, l/l_m,$$

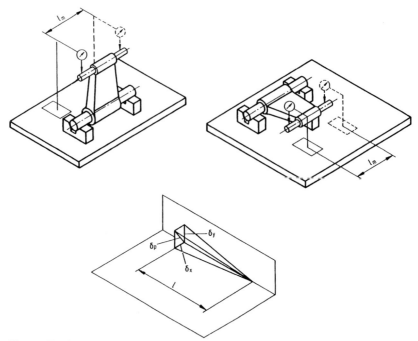

Figure 18.13: Measurement of the radial deviation of parallelism by detecting the deviations in rectangular coordinates.

where δ_u is the geometrical deviation, δ_m is the measured deviation over the length l_m, l is the length of the feature and l_m is the distance between the measured points.

The geometrical tolerances t are defined as widths of tolerance zones within which the toleranced feature must be contained. Geometrical deviations δ, however, are measurable deviations from the geometrical ideal form δ_f, orientation δ_d or location δ_o.

The maximum permissible deviation from location δ_o (positional deviation δ_c, coaxiality deviation δ_a, symmetry deviation δ_s) corresponds to one-half of the value of the location tolerances, $\delta_o = t_o/2$ (Figures 18.52 and 18.53; compare Figures 18.56 and 18.57 with Figure 3.4). The same applies to deviations of the profile of a line, δ_b or of a surface, δ_h, defined by theoretical exact (rectangular framed) dimensions (Figures 4.1–4.3). See Table 18.1.

In all remaining cases the maximum permissible geometrical deviation corresponds to the full value of the geometrical tolerance (Figures 8.28 and 18.48–18.51). See Table 18.1.

Table 18.1: Synopsis of geometrical deviations δ and comparison with geometrical tolerances t.

Charac-teristic		Reference ——— and toleranced element ———	
		with intersection	without intersection
Form	⌒ — ○ ⌓ ⌗ ⬭	$A_{p\,max}$ $A_{v\,max}$ $A_{p\,max}$ $A_{v\,max}$ $t \geq \delta = A_{p\,max} + A_{v\,max}$	$A_{\overline{min}}$ $A_{\overline{max}}$ t A_{min} A_{max} t $t \geq \delta = A_{max} - A_{min}$
Orientation	∠ // ⊥	$A_{p\,max}$ $A_{v\,max}$ t $t \geq \delta = A_{p\,max} + A_{v\,max}$	A_{max} A_{min} t $t \geq \delta = A_{max} - A_{min}$
Location	⊕ ◎ ≡ ⌒ ⌓	A_{max} t $t \geq 2\delta = 2A_{max}$	A_{max} t $t \geq 2\delta = 2A_{max}$
Run-out	↗ ↗↗		A_{min} A_{max} t $t \geq \delta = A_{max} - A_{min}$

18.7 METHODS OF INSPECTION

The following provides a survey of the most relevant inspection methods. The former East German Standards TGL 39 093 to TGL 39 098 and TGL 43 041 to

TGL 43 045 and TGL 43 529, TGL 43 530 provide more detailed descriptions of the same inspection methods. However, it should be noted that tolerances of orientation and location according to TGL 19 080 and according to the Comecon Standard ST RGW 301-76, in contrast to ISO 1101, are defined by eliminating the form deviations of the toleranced feature, i.e. deviations of orientation and location do not include deviations of form (Figure 18.14).

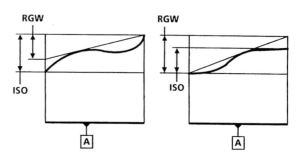

ISO : ISO 1101

RGW : ST RGW 301–76,   ST RWG 368–76
TGL 19 080,   TGL 31 049

Figure 18.14:   Deviation (tolerance) of orientation and location according to ISO 1101 and according to ST RGW 301-76 and ST RGW 368-76.

In the following the principles of the inspection methods corresponding to ISO 1101 are described.

18.7.1   Assessment of straightness deviations of lines of surfaces

18.7.1.1   Definition

The deviations of the line of the workpiece surface from an (almost) geometrical ideal reference straight line (embodiment, e.g. established by a straight edge) are measured. The line of the workpiece surface and the reference line are contained in a section plane (Figure 3.15), or a projection plane (i.e. a plane onto which the lines and the two parallel planes establishing the tolerance zone are projected) (Figures 3.19 and 3.22).

When the reference line does not intersect (but eventually touches) the workpiece line, the straightness deviation δ_g according to ISO 1101 is the difference between the largest and smallest distances between the workpiece line and the reference line (Figure 18.16 and Table 18.1).

When the reference line does intersect the workpiece line, the straightness deviation according to ISO 1101 is the sum of the largest distances of the workpiece line from the reference line on both sides of the reference line (Table 18.1), i.e. the range of the local deviations of the workpiece line from the reference line (Figure 8.28).

The reference line should be aligned according to the minimum requirement (Figure 18.1).

The straightness deviation δ_g must not exceed the straightness tolerance t_g:

$\delta_g \leq t_g$.

18.7.1.2 Type of detection

The deviations can be detected by

(1) continuously probing and recording;

(2) consecutive probing (sampling, approximation) and recording

while

(a) measuring the distance to the reference line (embodiment);

(b) measuring the inclination of a two point bridge on the surface relative to the reference line (embodiment)

18.7.1.3 Measurement methods

The reference line (embodiment) may be established by

- straightness guidance   (of a form measuring instrument)
  (of a coordinate measuring instrument);

- measuring plate   (with length measuring instrument, e.g. dial gauge)
  (optical flat with evaluating of contour lines);

- straight edge   (with length measuring device, e.g. dial gauge or feeler gauge);

- measuring wire   (with measuring microscope);

- straight line marking   (within profile projector);

- optical axis
  (collimating microtelescope with targets)*
  (autocollimator and test mirror)†
  (laser beam with photoelectric detector)*
  (laser interferometer with two-point bridge);†

- earth curvature
  (liquid surface of a hose levelling instrument)*‡
  (two-point bridge with inclinometer).†‡

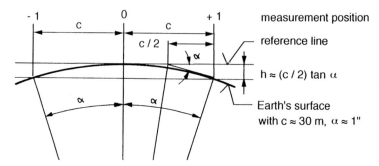

Figure 18.15: Height correction *h* over a distance *c* for eliminating the effect of the Earth's curvature.

Table 18.2: Height correction *h* over a distance *c* for eliminating the effect of the Earth's curvature (see Figure 18.15).

c (m)	h (μm)	c (m)	h (μm)	c (m)	h (μm)
1	0.08	11	9.8	25	50
2	0.32	12	11	30	73
3	0.73	13	14	35	99
4	1.3	14	16	40	129
5	2.0	15	18	45	160
6	2.9	16	21	50	203
7	3.9	17	23	60	291
8	5.2	18	26	70	396
9	6.5	19	29	80	517
10	8.0	20	32	100	800

---

*Height measurement.
†Inclination measurement.
‡With long reference lines the curvature of the Earth ($\approx 1''$ per 30 m) must eventually be taken into account (Figure 18.15, Table 18.2).

When the surface (workpiece line) is concave, the reference line (embodiment) aligns automatically according to the minimum requirement (Figure 18.16).

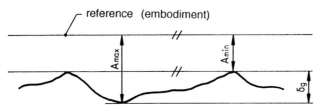

Figure 18.16: Measurement of straightness deviation; alignment according to the minimum requirement.

When the surface (workpiece line) is convex, the reference line is to be aligned so that both deepest points of the workpiece line have equal distance from the reference line. An approximation is that both ends have equal distance from the reference line (Figure 18.17).

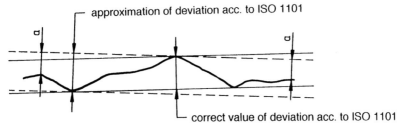

Figure 18.17: Measurement of straightness deviation; alignment according to the minimum requirement: continuous lines, exact; dashed lines, approximation by equal distance at both ends.

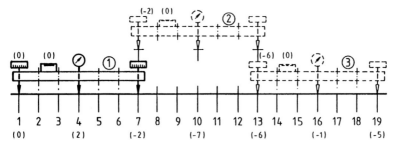

Figure 18.18: Consecutive measurement of straightness deviation with adjusted straight edge without overlapping, example with measured values in parentheses, according to TGL 39 093.

207

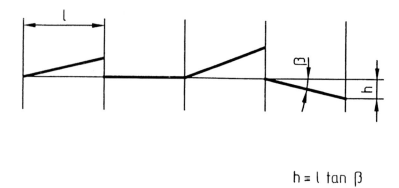

$$h = l \tan \beta$$

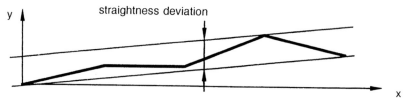

straightness deviation

Figure 18.19:   Graphical evaluation of straightness deviation from consecutive measurement of the deviation from the horizontal orientation.

When form measuring instruments or coordinate measuring machines are used, the alignment, as an approximation, is often performed according to the Gaussian regression line (sum of squares of the distances to the regression line is a miminum). Then the obtained value of the straightness deviation can be larger than the value according to ISO 1101 (minimum zone value).

When the toleranced length is longer than the straightness embodiment the embodiment, e.g. a straight edge with height-adjustable supports, is used in consecutive measuring positions without (a) or with (b) overlapping.

(a) When the straightness embodiment is used in positions without overlapping, it is to be adjusted according to a levelling instrument. The first height indication of the following position is to be adjusted according to the last height indication of the preceding embodiment position (Figure 18.18).

The adjustment of the connecting points of the straight edge during measurement can be omitted and be transferred into the graphical evaluation (Figure 18.19).

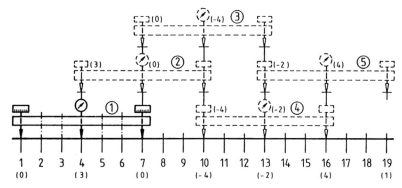

Figure 18.20: Measurement of straightness deviation with overlapping straight edges, example with measured values in parentheses, according to TGL 39 093.

(b) When the straightness embodiment is used in overlapping positions, at least two measurement positions must overlap. The two overlapping height indications of the following position have to be adjusted according to the height indications of the preceding embodiment position (alignment of the straight edge) (Figure 18.20 and 18.21). The measurement uncertainty is lower the more the measurement positions overlap.

When the angle α (inclination) of the connection of adjacent measuring positions relative to the reference line (straightness embodiment) is measured in seconds, the height difference is $h = c \tan \alpha$ in μm, where c the distance in m between the measuring positions ($\tan 1'' = 4.848 \, \mu\text{m}/1000 \, \text{mm} \approx 4 \, \mu\text{m}/1000 \, \text{mm}$) (Figure 18.22).

18.7.2 Assessment of straightness deviations of axes

18.7.2.1 Definition

With cylindrical features, the actual axis can be taken as a sequence of centres of circles that can be defined as follows:

(a) regression circle (by Gauss, the circle of which the sum of the squares of the radial distances to the circumference is a minimum, namely the least squares circle or LSC);

(b) minimum-zone circle (by Chebychev, the circle of which the maximum radial distance to the circumference is a minimum, namely the MZC);

(c) contacting circle (the maximum inscribed circle for holes, MIC and the

209

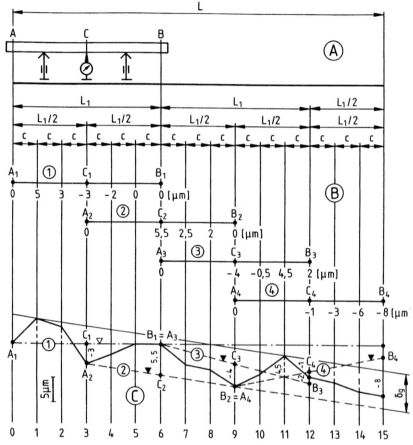

A measuring scheme and length of measuring steps
B positions of straight edge and examples of distances to the straight edge related to position A of the straight edge
C examples of graphical evaluation of the measuring result

Figure 18.21: Graphical evaluation of the straightness deviation for consecutive measurement of the deviations using overlapping positions of the straight edge, example with measured values, according to TGL 39 093.

minimum circumscribed circle for shafts, MCC, sometimes referred to as the plug gauge circle, PGC, and ring gauge circle, RGC);

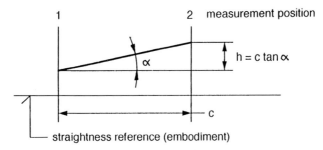

Figure 18.22:   Inclination α along a distance c converted to height deviation h.

in cross sections perpendicular to the axes of the following reference cylinders:

(a) regression cylinder (by Gauss, the cylinder of which the sum of the squares of the radial distances to the actual cylinder surface is a minimum, namely the least-squares cylinder);

(b) minimum-zone cylinder (by Chebychev, the cylinder of which the maximum radial distance to the actual cylinder surface is a minimum);

(c) contacting cylinder (the maximum inscribed cylinder for holes and the minimum circumscribed cylinder for shafts);

(d) two cross-section cylinder, defined by the regression circles or the minimum-zone circles or the contacting circles of two cross-sections near the ends of the feature.

According to the ISO Standards, it is not yet standardized which definition in which cases applies. Therefore this must be agreed between the parties.

Coordinate measuring machines apply normally to the regression cylinder and the regression circle and, if optionally available, the minimum-zone circle and the contacting circles. Form measuring instruments apply normally to the two cross-section cylinder defined by the regression circles or minimum-zone circles or contacting circles. (But so far with coordinate measuring machines the reference cylinder is always the regression cylinder, because the mathematical techniques have been developed only for this, while with form measuring instruments it is normally the two cross-sections cylinder).

The regression circle and the regression cylinder are the only methods that are always unique and that need the least number of measurements (probes, measured points) to be sufficiently stable in size and in location.

The deviations of the actual axis from an (almost) geometrical ideal reference

Inspection of Geometrical Deviations

straight line (embodiment) are measured. The deviations have to be calculated in relation to an (imagined) reference line to which the minimum requirement applies, i.e. the maximum local deviation e must be a minimum (Figures 18.1 and 18.24). The straightness deviation δ_g is the range of the local deviations e: $\delta_g = 2e_{max}$ (Figures 8.28 and 18.24). The straightness deviation must not exceed the straightness tolerance t_g, i.e. the maximum local straightness deviation e_{max} must not exeed one-half of the straightness tolerance t_g:

$$\delta_g = 2e_{max} \le t_g.$$

18.7.2.2 Assessment of coordinates of axes

Depending on the drawing indication, the straightness tolerance is defined as a zone in a projected plane (also expressed as the distance between two parallel planes) in order to assess the curvature projected in this plane (Figure 3.22) or as a cylindrical zone in order to assess the curvature in space (Figure 3.17). The evaluation of the deviation must be performed appropriately in each case.

With tolerance zones in projected planes or between parallel planes, the evaluation of the deviation is similar to that described in 18.7.1.

A cylindrical tolerance zone, i.e. one for the assessment of the straightness deviation (curvature) in space, is normally measured in each cross-section in two mutually perpendicular directions (coordinates x_i, y_i, z_i, Figure 18.23).

The smallest cylinder containing all assessed centres, is to be determined. There are several approximations for this purpose.

One is to define a reference line (z direction), to align the workpiece in this coordinate system, to assess the actual axis (coordinates) and to calculate the distances between the coordinates of the actual axis and the reference line. The local distance in space is

$$e_i = \sqrt{(x_i^2 + y_i^2)},$$

where x_i is the distance in the xz plane (x direction) and y_i is the distance in the yz plane (y direction). The diameter of the smallest cylinder containing all centres (points of the actual axis) is $\delta_g = 2e_{max} \le t_g$. Its axis is the reference line (Figure 18.23).

Often the reference line is established by the centres of the first and last cross-sections. With coordinate measuring machines, the reference line is the axis of the substitute cylinder (regression cylinder according to the Gaussion method) (see 8, 18.7.2.1 and 18.3.3).

When the reference line is not exactly parallel to the z axis, but only approximately so, the measured coordinates x_i can be recorded in an xz coordinate system and y_i in a yz coordinate system (Figure 18.24). For an approximation in each coordinate system the reference line is to be determined so that the maximum deviation is a minimum (parallel to the two parallel straight lines of minimum

212

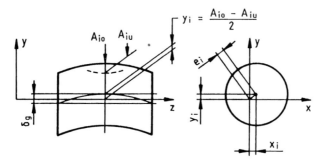

Figure 18.23: Assessment of coordinates of an actual axis (centres of cross-sections).

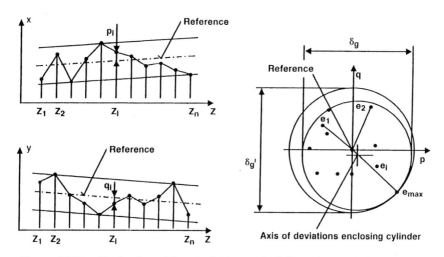

Figure 18.24: Evaluation of the straightness deviation of an axis in space.

distance that enclose all points of the actual axis; see 18.7.1.3). The deviations p and q of the points from the reference line are to be determined. The local actual deviations in space from the reference line are $e_i = \sqrt{(p_i^2 + q_i^2)}$, and the diameter of the smallest cylinder that contains all centres e_i (points of the actual axis) is the straightness deviation δ_g. The straightness deviation δ_g must not exceed the straightness tolerance t_g: $\delta_g \leq t_g$.

An approximation, easy to calculate, is (Figure 18.24)

$$\delta_g \approx \delta_g' = 2e_{max} \leq t_g.$$

213

18.7.2.3 Assessment with one dial indicator

A simple approximate method for assessing the straightness deviation of an axis is shown in Figure 18.25. The workpiece has to be supported centrally (e.g. in the centres of the measuring positions z_1 and z_n.

In each cross-section (measurement position) the coordinates of the centres are to be determined by probing at the angle positions 0° and 180°, 90° and 270°. The coordinates of the centre (observing the signs + and −) are to be introduced in a polar diagram (in the x direction $x_i = (A_0 - A_{180})/2$ and in the y direction $y_i = (A_{90} - A_{270})/2$ (Figure 18.23). The diameter of the smallest circle enclosing the centres of all cross-sections corresponds to the straightness deviation δ_g.

With this method, problems arise with positioning the axis of revolution (reference line) according to the minimum requirement and with the assessment of the actual centres according to the definition given in 18.7.2.1.

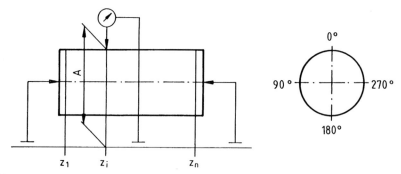

Figure 18.25: Assessment of straightness deviation of an axis with one dial indicator.

18.7.2.4 Assessment with two dial indicators

Another simple approximation method for assessing the straightness deviation of an axis is shown in Figure 18.26. The workpiece axis has to be aligned parallel to the reference line (embodiment, e.g. a measuring table), for example through the centres of the measuring positions z_1 and z_n with the same distance from the measuring table.

In each longitudinal section (containing the axis), of e.g. 4 sections, the values $R = (A_0 - A_u)/2$ are to be determined at several (at least three) measuring positions. The difference between R_{max} and R_{min} within one section represents the straightness deviation of the axis in this section. The straightness deviations of the axis of the cylindrical feature is the maximum of the straightness deviations of the sections (Figure 18.27):

214

straightness deviation of the axis $\delta_g = \left(\dfrac{A_0 - A_u}{2}\right)_{max} - \left(\dfrac{A_0 - A_u}{2}\right)_{min} \leq t_g$.

As the directions of measurement are opposite, it is the difference of half the sums of the distances from the measuring plate that is determined.

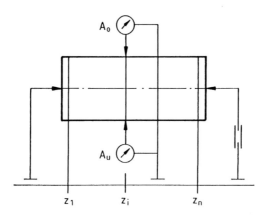

Figure 18.26: Assessment of straightness deviation of an axis with two dial indicators.

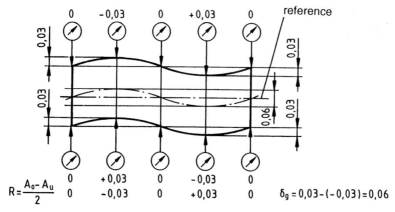

Figure 18.27: Example of assessment of the straightness deviation of an axis with two dial indicators.

18.7.2.5   Assessment with form measuring instrument or coordinate measuring machine

The instruments assess the coordinates of points of the actual circumference in cross-sections approximately perpendicular to the axis and calculate the coordinates of the actual centre points of which the actual axis is composed. From the points of the actual axis, they calculate the local straightness deviations, taking into account the minimum requirement. Twice the value of the maximum local straightness deviation corresponds to the straightness deviation δ_g according to ISO 1101 (see 18.7.2.2).

The instruments allow measurements close to the definitions. However, the definition of the actual axis has to be agreed upon (see 18.7.2.1).

18.7.3   Assessment of flatness deviations

18.7.3.1   Definition

The deviations of the surface from an (almost) ideal reference plane (embodiment, e.g. established by a measuring plate) are measured.

When the reference plane does not intersect (but eventually touches) the workpiece surface, the flatness deviation δ_e according to ISO 1101 is the difference between the largest and smallest distances between the workpiece surface and the reference plane (Table 18.1).

When the reference plane does intersect the workpiece surface, the flatness deviation δ_e according to ISO 1101 is the sum of the largest distances between the reference plane and the workpiece surface above and below the reference plane (Table 18.1).

The reference plane is to be aligned according to the minimum requirement (Figure 18.1). The flatness deviation δ_e must not exceed the flatness tolerance t_e: $\delta_e \leq t_e$.

18.7.3.2   Types of detection and measurement methods

These are similar to those used for the assessment of straightness deviations (see 18.7.1.2 and 18.7.1.3).

When the deviations are assessed in lines with the aid of straightness embodiments (e.g. with a straight edge), the embodiment must remain in the same plane (inclination, height level), or the measured values must accordingly be corrected by calculation.

Figure 18.28 shows a measurement device where the alignment is controlled by an inclination measuring instrument.

However, it is difficult to align the workpiece surface relative to the plane embodiment so that the largest measured deviation is a minimum. Often, as an

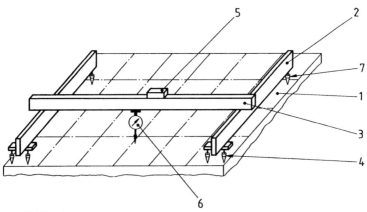

Figure 18.28: Assessment of flatness deviation with straight edge, inclinometer and dial gauge.

approximation, the distances between workpiece surface and plane embodiment are equalized at three ends (points) of the surface (e.g. three supports of equal height). Another approximation, when a computer is used (e.g. with coordinate measuring machines) is alignment parallel to the regression plane (least-squares plane, the plane of which the sum of the squares of the distances to the actual surface is a minimum). With the approximations, there are always larger estimations of the deviations than from the minimum requirement according to ISO 1101.

The plane embodiment according to the minimum requirement has one of the following positions (alignments)

(a) it touches the three highest points (concave form);

(b) it touches the three deepest points (convex form);

(c) it touches the two highst points and is parallel to the straight line touching the two deepest points (saddle form, Figure 18.29).

The former East German Standard TGL 39 094 describes a graphical method "ZNIITM" for the assessment of the flatness deviation close to the definition (minimum requirement).

The distances of the workpiece surface from the plane embodiment are measured. The inclination of the embodiment relative to the workpiece surface should be small.

The measurement positions (points) are enumerated and plotted on a scale, A in Figure 18.30, measuring points A1 to C5. At each point the measured value is indicated. In the view B are plotted on the scale the height differences from the plane embodiment. The straight line P_x touches the points from above and is aligned with respect to the points according to the minimum requirement. P_x determines one direction of the reference plane.

In an inclined plane C, perpendicular to the straight line P_x (connection B1–A4), the distances of the surface points from P_x (in view B) are plotted (e.g. a). The straight line P_y touches the highest point and is directed according to the minimum requirement ($\delta_e = \min$). P_y determines the other direction of the reference plane. The point with the largest distance (in the direction of P_x, B1–A4) to P_y corresponds to the flatness deviation δ_e. (The reference plane in Figure 18.30 is determined by the points B1, C3 and A4).

Another method according to TGL 39 094 for the assessment of the flatness deviation close to the definition uses analogue mechanical devices (Figure 18.31).

The pins are movable and to be aligned on a larger scale according to the measured deviations. As all pins are of equal length, the other side exhibits a mirror image of the measured form. By inclining the device on a measuring plate, the position is to be determined in which the pin with the largest distance to the measuring plate has the least distance. This distance corresponds to the flatness deviation δ_e.

18.7.3.3 Assessment of flatness deviation with straight edge and dial indicator

The procedure of the assessment of the flatness deviation with straight edge and dial indicator (Figure 18.32) is as follows:

(1) straight edge in position C1–A5, supports adjusted to 0;

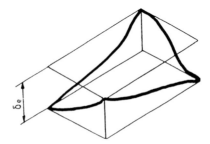

Figure 18.29: Flatness deviation δ_e of saddle form surface.

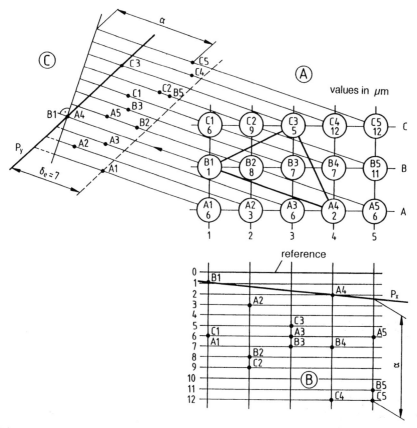

Figure 18.30: Assessment of flatness deviation according to the method ZNIITM.

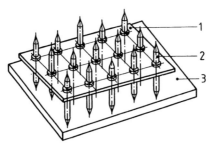

Figure 18.31: Analogue mechanical device for the assessment of the flatness deviation; 1, plate; 2, adjustable pins; 3, workpiece/measuring plate.

(2) centre point B3 measured and registered;

(3) straight edge in position A1–C5, value of B3 from step 2 aligned and straight
edge at the ends aligned to equal distances;
the two diagonals define the plane embodiment; measurements A1 and C5
registered;

(4) straight edge in position A1–C1 aligned to the already registered measured
values A1 and C1, B1 measured and registered etc.

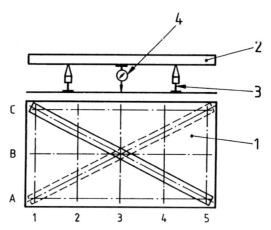

Figure 18.32:   Assessment of flatness deviation with straight edge and dial
indicator: 1, workpiece; 2, straight edge; 3, adjustable supports;
4, dial gauge.

18.7.4   Assessment of roundness deviations

18.7.4.1   Definition

The deviations of the workpiece circumferece from an (almost) ideal reference
circle (embodiment, e.g. established by a circular movement) are measured in
cross-sections perpendicular to the axis (see 18.3.2 and 18.3.3).

When the reference circle does not intersect (but eventuelly touches) the
workpiece circumference line, the roundness deviation δ_r according to ISO 1101
is the difference between the largest and smallest radial distances of the workpiece
circumference from the reference circle (Figure 18.33 and Table 18.1).

When the reference circle does intersect the workpiece circumference line the
roundness deviation δ_r according to ISO 1101 is the sum of the largest radial

distances of the workpiece circumference line from the reference circle on both sides of the reference circle (Table 18.1), i.e. the range of the local deviations e of the workpiece circumference line from the reference circle (Figure 18.34).

The reference circle is to be aligned according to the minimum requirement (Figure 18.2). See also 18.3.2 and 18.3.3.

The roundness deviation δ_r must not exceed the roundness tolerance t_r: $\delta_r \leq t_r$.

For the assessment of the roundness deviation with coordinate measuring machines or form measuring instruments the following reference circles are standardized according to ISO 4291:

(a) regression circle (LSC), basis for tolerance or deviation zone Z_q;

(b) minimum zone circle (MZC), basis for tolerance or deviation zone Z_z;

(c) contacting circle (MIC, MCC), basis for tolerance or deviation zone Z_i or Z_c.

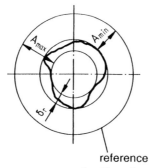

reference

Figure 18.33: Roundness measurement with reference circle that does not intersect the workpiece circumference line. $\delta_r = A_{max} - A_{min} \leq t_r$.

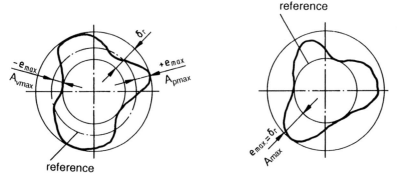

reference

Figure 18.34: Roundness measurement with reference circles that intersect or touch the workpiece circumference line. $\delta_r = A_{pmax} + A_{vmax} \leq t_r$.

See 18.7.2.1. These reference circles intersect or touch the workpiece circumference line.

According to ISO 1101 the minimum-zone circle (minimum requirement) applies. However, for practical reasons sometimes the other definitions are applied.

The minimum-zone circle leads to the smallest values of the roundness deviation. According to TGL 39 096, the (random) differences in the values of the roundness deviation caused by the different definitions of the reference circle are up to +15%.

The regression circle needs fewer measurement points than the other reference circles to be sufficiently stable. Therefore the regression circle is prefered in the measurement technique.

For the definition of the cross-sections perpendicular to the axis see 18.7.2.1 and 18.3.3.

18.7.4.2 Measurement methods

The reference circle (embodiment) may be established by

- high precision circular movement   (form measuring instrument) (measurement while workpiece mounted in centres or in a chuck);

- high-precision straightness guidances perpendicular to each other and calculation of circles   (coordinate measuring machines);

- revolving on a measuring plate   (measurement of diameters);

- revolving in a V-block   (three-point measurement);

- circular device   (measurement in a ring or on a plug);

- circular marking   (profile projector).

When polar diagrams are used, it should be noted that the form in the profile diagram looks quite different than the form of the real profile because of the amplification of radial distances (Figure 18.35).

18.7.4.3 Two- and three-point measurements

For the methods see Figures 18.36 and 18.37.

With two-point measurements (measurements of diameter), lobed forms cannot be detected (Figure 18.36). When lobed forms can occur (e.g. with centre-less grinding or reaming), in addition to the two-point measurement other inspections (e.g. three-point measurements, Figure 18.37) are necessary.

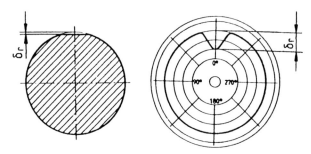

Figure 18.35: Distortion of a profile diagram caused by amplification in the diagram: left, without amplification and without distortion; right, with amplification and distortion.

With three-point measurements (e.g. in V-blocks, Figure 18.37), the possibility of detecting the different types of lobed forms and other forms differs depending on the type of form deviation (e.g. type of lobed form) and on the type of measurement method. Table 18.3 gives correction values k. The measured values $\Delta = A_{max} - A_{min}$ are to be divided by k in order to correspond to the roundness deviation δ_r:

$$\delta_r = \Delta/k = (A_{max} - A_{min})/k \leq t_r.$$

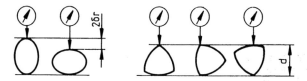

Figure 18.36: Oval and lobed forms.

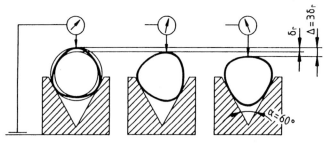

Figure 18.37: Assessment of form deviations of lobed forms in V-blocks: summit method.

223

With three-point measurements a distinction must be made between the summit method (V-support, two fixed anvils on one side and the measuring anvil, indicator, on the other side of the workpiece; Figures 18.37–18.39) and the rider method (two fixed anvils and the measuring anvil, indicator, on the same side of the workpiece, Figure 18.40).

Table 18.3: Correction values k for the measurement of form deviations by two-point measurements and three-point measurements: n is the order of the harmonics (number of undulations) to be assessed

	Angles α and α/β of measurement methods (Figures 18.38–18.40)												
	Summit method								Ryder method				
n	60°	72°	90°	108°	120°	180°	60°/30°	120°/60°	60°	72°	90°	108°	120°
2	–	0.47	1	1.4	1.6	2	1.4	2.4	2	1.5	1	0.62	0.42
3	3	2.6	2	1.4	1	–	2	2	3	2.6	2	1.4	1
4	–	0.38	0.41	–	0.42	2	1.4	1	2	2.4	2.4	2	1.6
5	–	1	2	2.2	2	–	2	2	–	1	2	2.2	2
6	3	2.4	1	–	–	2	0.73	0.42	1	0.38	1	2	2
7	–	0.62	–	1.4	2	–	2	2	–	0.62	–	1.4	2
8	–	1.5	2.4	1.4	0.42	2	1.4	1	2	0.47	0.4	0.62	1.6
9	3	2	–	–	1	–	2	2	3	2	–	–	1
10	–	0.70	1	2.2	1.6	2	1.4	2.4	2	2.7	1	0.24	0.42
11	–	2	2	–	–	–	–	–	–	2	2	–	–
12	3	1.5	0.41	0.38	2	2	0.73	1	1	0.47	2.4	0.62	–
13	–	0.62	2	1.4	–	–	–	–	–	0.62	2	1.4	–
14	–	2.4	1	–	1.6	2	1.4	0.42	2	0.38	1	2	0.42
15	3	1	–	2.2	1	–	2	2	3	1	–	2.2	1
16	–	0.38	2.4	–	0.42	2	1.4	1	2	2.4	0.41	2	1.6
17	–	2.6	–	1.4	2	–	2	2	–	2.6	–	1.4	2
18	3	0.47	1	1.4	–	2	0.73	2.4	1	1.5	1	0.62	2
19	–	–	2	–	2	–	2	2	–	–	2	–	2
20	–	2.7	0.41	2.2	0.42	2	1.4	1	2	0.7	2.4	0.24	1.6
21	3	–	2	–	1	–	2	2	3	–	2	–	1
22	–	0.47	1	1.4	1.6	2	1.4	0.42	4	1.5	1	0.62	0.42

–, the method gives no indication of the roundness deviation. 180° is a two-point measurement.

The number of sides n of the lobed form (number of undulations) may be detected by counting the maxima (or minima) during one revolution of the workpiece in the V-block (e.g. in the 72° V-block). When there is a superposition of harmonics (undulations) (see Figures 18.111 and 18.113), selection of the proper correction value is practically impossible. The correction value must be estimated. In this case TGL 39 096 recommends the asymmetrical summit method $\alpha = 120°/\beta = 10°$ and the average correction value 1.2.

In order to cover all possible form deviations and numbers of undulations, BS

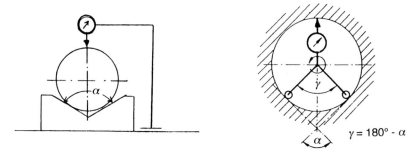

Figure 18.38: Three-point measurement: summit method, symmetrical setting.

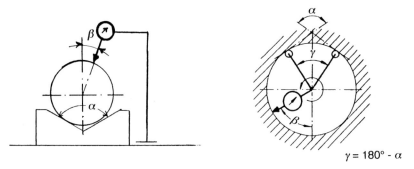

Figure 18.39: Three-point measurement: summit method, asymmetrical setting.

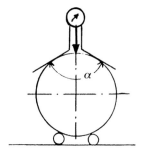

Figure 18.40: Three-point measurement: rider method, symmetrical setting.

225

Inspection of Geometrical Deviations

3730 : Part 3 : 1982 recommends that one two-point measurement and two three-point measurements be taken at different angles between fixed anvils and the angles be selected from the following:

symmetrical setting $\alpha = 90°$ and $120°$ or $\alpha = 72°$ and $108°$;

asymmetrical setting $\alpha = 120°$, $\beta = 60°$ or $\alpha = 60°$, $\beta = 30°$;

where α is the angle between fixed anvils and β is the angle between the direction of measurement and the bisector of the angle between fixed anvils (Figure 18.39).

The measuring anvil should be selected from Table 18.4. ISO 4292 gives a similar table, but does not specify spherical or cylindrical anvils.

Table 18.4: Types of anvil

Surface form	Anvil radius (mm)	Surface radius (mm)
Convex surface	Spherical: 2.5	All
Convex edge	Cylindrical: 2.5	All
Concave surface	Spherical: 2.5	≥ 10
Concave edge	Cylindrical: 2.5	≥ 10
Concave surface	Spherical: 0.5	< 10
Concave edge	Cylindrical: 0.5	< 10

It is further recommended that the following fixed anvils be used:

• for external measurement, a V-support with a small radius; the median plane of the V-support should be in the same plane as the plane of measurement;

• for internal measurement, a sphere with a small radius; the median plane of the sphere should be in the same plane as the plane of measurement.

With a combination of two-point measurement, three-point measurement and evaluation by calculation according to U. Barth (see Ref. [10]) (Figure 18.41), approximations to the smallest diameter, the mating size and the roundness deviation can be assessed. This method can assess form deviations up to the 10th harmonic (e.g. the number of sides of the lobed form). It is not necessary to know the number of the harmonic.

The actual sizes are detected by the dial indicator A_2. The mating size P and the roundness deviation δ_r^* are to be calculated as follows.

When the indicator reading differences $\Delta A_1 \leq A_2$

$$P \approx A_{2min} + \Delta A_2 = A_{2max},$$

---

*Applicable to disc-shaped workpieces or when the straightness and parallelism deviations are negligable.

226

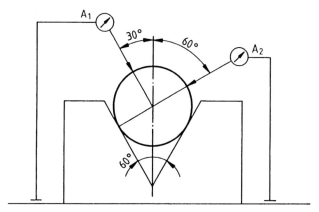

Figure 18.41: Combination of two- and three-point measurements according to U. Barth for the determination of the smallest actual size, the mating size and the roundness deviation.

$$\delta_r \approx \Delta A_2/2 \leq t_r.$$

When the indicator reading differences $\Delta A_1 > \Delta A_2$

$$P \approx A_{2min} + \Delta A_1/2,$$
$$\delta_r \approx \Delta A_1/2 \leq t_r.$$

18.7.4.4 Assessment with measuring ring or plug

Figure 18.42 shows a measuring ring according to the former East German Standard TGL 39 096. The inner ring surface establishes the circle embodiment. The diameter is adjustable within the range of the size tolerance.

For assessment of the roundness deviations of holes a similar method with measuring plugs can be used.

These assessments are close to the definition of the roundness deviation according to ISO 1101.

18.7.5 Assessment of cylindricity deviations

18.7.5.1 Definition

The deviations of the workpiece surface from an almost-ideal reference cylinder (embodiment, e.g. established by circular movements and movements perpendicular to the plane of the circular movements) are measured.

When the reference cylinder does not intersect the workpiece surface (but

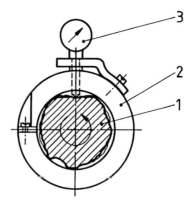

Figure 18.42: Assessment of roundness deviations with a measuring ring: 1, workpiece; 2, embodiment (ring); 3, dial guage.

eventually touches it), the cylindricity deviation δ_z according to ISO 1101 is the difference between the largest and smallest radial distances of the workpiece surface from the reference cylinder (Table 18.1). Similarly to Figure 18.33, it is

$$\delta_z = A_{max} - A_{min}.$$

When the reference cylinder does intersect the workpiece surface, the cylindricity deviation according to ISO 1101 is the sum of the largest radial distances of the workpiece surface from the reference cylinder on both sides of the reference cylinder (Table 18.1), i.e. the range of the local deviations e of the workpiece surface from the reference cylinder (Figure 18.34).

The reference cylinder is to be aligned according to the minimum requirement (Figure 18.2). When methods (a), (b), (c) and (e) of 18.7.5.2 (measurement in sections) are used, the alignment of the workpiece according to 18.3.2 must be observed (see also 18.3.3).

The cylindricity deviation δ_z must not exceed the cylindricity tolerance t_z: $\delta_z \leq t_r$.

For the assessment of the cylindricity deviation with coordinate measuring machines or form measuring instruments the following reference cylinders are used:

(a) regression cylinder (by Gauss, the cylinder of which the sum of the squares of the radial distances to the surface is a minimum, namely the least-squares cylinder);

(b) minimum-zone cylinder (by Chebychev, the cylinder of which the maximum radial distance to the surface is a minimum);

(c) contacting cylinder (the maximum inscribed cylinder for holes and the minimum circumscribed cylinder for shafts).

These reference cylinders intersect or touch the workpiece surface. According to ISO 1101, the minimum-zone cylinder (minimum requirement) applies. However, for practical reasons sometimes the other definitions are applied.

The minimum-zone cylinder leads to the smallest values of cylindricity deviation δ_z.

The regression cylinder needs fewer measurement points than the other reference cylinders to be sufficiently stable. Therefore it is preferred in the measurement technique.

18.7.5.2 Measurement strategies

The following measurement strategies should be distinguished:

(a) radial-section method;

(b) generatrix method;

(c) generatrix and radial section method;

(d) helical method;

(e) extreme-positions method;

(f) point method.

In all methods the measured profiles or points must be related to the same coordinate system. The workpiece (feature) axis has to be aligned parallel to the straight line guidance of the measuring device. This alignment can also be replaced by calculation. For the definition of the axis see 18.7.2.1. See also 18.3.3.

(a) Radial-section method The profile lines (circumferences) of several cross-sections perpendicular to the reference cylinder (axis of measurement) are plotted in one common polar diagram and evaluated according to the minimum requirement (Figure 18.43).

(b) Generatrix method In several sections containing the reference cylinder axis (axis of measurement) the two always opposite profile lines (generatrixes) are assessed, plotted in one common diagram and evaluated according to the minimum requirement (Figure 18.44).

(c) Generatrix and radial section method This is a combination of the generatrix method and the radial section method. It gives the smallest measuring uncertainty

because in both directions (axial and radial) the deviations are assessed with small sampling intervals (e.g. according to the Nyquist theorem), see 18.11.1.

(d) Helical method The profile lines assessed by a helical probing trace, probing perpendicular to the reference cylinder axis (axis of measurement), are plotted in one common diagram and evaluated according to the minimum requirement (Figure 18.45). Eventually the profiles of two cross-sections at the features end perpendicular to the reference cylinder (axis of measurement) are also assessed and included in the diagram and evaluation.

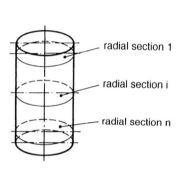

radial section 1

radial section i

radial section n

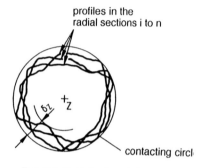

profiles in the radial sections i to n

contacting circl

Z centre of contacting circle

Figure 18.43: Radial-section method.

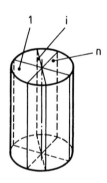

1 i

n

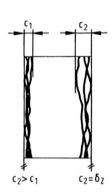

c_1 c_2

$c_2 > c_1$ $c_2 = \delta_z$

Figure 18.44: Generatrix method.

(e) Point method Random distributed points of the workpiece surface are assessed, plotted in one common polar diagram perpendicular to the reference cylinder (axis) and evaluated according to the minimum requirement (Figure 18.46).

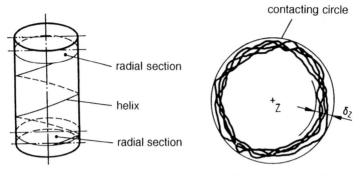

contacting circle

radial section

helix

radial section

$+Z$

δ_Z

Z centre of contacting circle

Figure 18.45:   Helical method.

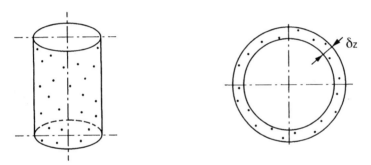

δz

Figure 18.46:   Point method.

(f) Extreme-positions method   According to TGL 39 097, the following are assessed and plotted:

● generatrix in one section containing the reference cylinder axis (generator lines at 0° and 180°);

● circumference lines in cross-sections perpendicular to the reference cylinder axis in the positions where the largest (max) and smallest (min) distances of the generator lines occur (cross-sections I and II).

The evaluation is shown in Figure 18.47. Z_1, c_{max} and Z_2, c_{min} in the section 0°–180° are the same in both Figure 18.47 top and bottom. The cylindricity deviation is $\delta_z = a_1 + a_2 + a_3 \leq t_z$.

231

Inspection of Geometrical Deviations

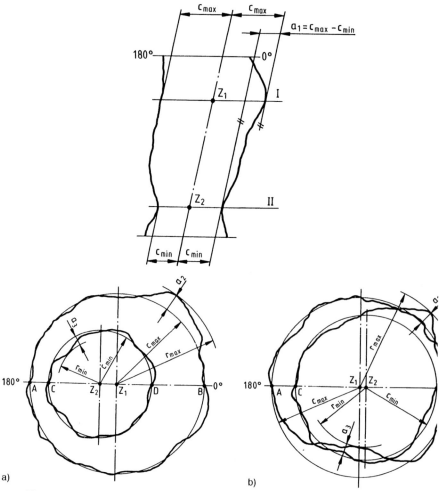

Figure 18.47: Extreme positions method, evaluation: (a) when the radial profiles do not intersect; (b) when they do.

18.7.5.3   Measurement methods

Measuring the cylindricity deviation   The reference cylinder (embodiment) may be established by

- high precision circular movement   (form measuring instrument)

232

and high precision straightness guidance perpendicular to the circular movement (measurement while workpiece is mounted in centres or in a chuck);

- high-precision straightness guidance perpendicular to each other and cylinder calculation (coordinate measuring machine);

- revolving in V-block and parallel straightness guidance (three-point measurement with V-block, straight edge, dial indicator and measuring table).

The three-point measurement does not allow to establish a stable workpiece coordinate system (the workpiece axis has no stable position in the coordinate system of measurement). Therefore this method is a rough approximation (see 18.7.4.3).

Assessment by measurement of components (a) Approximate evaluation of the cylindricity deviation δ_z from the roundness deviation δ_r and the parallelism deviation δ_p* of the generator lines:

$$\delta_z \approx \delta_r + \delta_p/2 \le t_z.$$

(b) Approximate evaluation of the cylindricity deviation δ_z from the roundness deviation δ_r and the longitudinal section profile deviation δ_q (see 21.4):

$$\delta_z \approx \delta_r + \delta_q \le t_z.$$

The approximation (b) with δ_q is considered to be closer to the correct assessment of the cylindricity deviation than the approximation (a) (with δ_p) (see Ref. [12]).

18.7.6 Assessment of profile deviations of lines

18.7.6.1 Definition

The deviations of the workpiece section (profile) line from an almost ideal reference line (embodiment) are measured in cross sections, the orientation of which is defined by the datum system (related profile tolerance) or perpendicular to the

---

*The cylindricity deviation comprises deviations of roundness, straightness and parallelism. As the parallelism deviation δ_p according to ISO 1101 comprises the straightness deviations and the inclinations of the generator lines relative to each other, the cylindricity deviation δ_z cannot be larger than the sum of the roundness deviation δ_r and the parallelism deviation δ_p. In most cases it is smaller (see Ref. [12]).

desired workpiece alignment direction (unrelated profile tolerance).
The desired workpiece alignment direction may be defined as

- perpendicular to a defined shoulder, determined by the regression plane or the minimum zone plane or the datum plane, see 18.3.3.1c)

- parallel to a defined workpiece established axis (of another feature) as described in 18.3.3.1.

The location of the reference line is defined relative to the datum system (related profile tolerance) or is defined by the minimum requirement, i.e. the maximum distance of the surface line from the reference line shall be a minimum (unrelated profile tolerance).

(Regarding the alignment of the workpiece in principle the similar effects occur as explained in 18.3.3.2 and 18.3.3.3.)

In the case of a related profile tolerance the profile deviation δ_b is the maximum distance of the workpiece section (profile) line from the reference line, measured perpendicular to the reference line (Table 18.1). The profile deviation δ_b must not exceed half of the profile tolerance t_b.

In the case of a unrelated profile tolerance the profile deviation δ_b is the sum of the largest distances of the workpiece section (profile) line from the reference line on both sides of the reference line (range), each measured perpendicular to the reference line (Table 18.1). The profile deviation δ_b must not exceed the profile tolerance t_b.

18.7.6.2 Measuring methods

The reference line (embodiment) may be established by:

- profile template   (copying system†)

- profile marking   (profile projector)

- high precision straightness   (coordinate measuring machine)
  guidances perpendicular to each
  other and calculation of the
  reference line

---

†The copying system probes the workpiece and the template with the same tip radius and records the deviations of the workpiece from the template.

18.7.7 Assessment of profile deviations of surfaces

18.7.7.1 Definition

The deviations of the workpiece surface from an almost ideal reference surface (embodiment, e.g. established by a form template or by calculation in a coordinate measuring machine) are measured.

For the alignment of the workpiece the same applies as to the assessment of profile deviations of lines, see 18.7.6.1.

In the case of a related profile tolerance the profile deviation δ_h is the maximum distance of the workpiece surface from the reference surface, measured perpendicular to the reference surface (Table 18.1). The profile deviation δ_h must not exceed half of the profile tolerance t_h.

In the case of a unrelated profile tolerance the profile deviation δ_h is the sum of the largest distances of the workpiece surface from the reference surface on both sides of the reference surface (range), each measured perpendicular to the reference surface (Table 18.1). The profile deviation δ_h must not exceed the profile tolerance t_h.

18.7.7.2 Measuring methods

The reference line (embodiment) may be established by:

- surface template   (copying system)

- high precision straightness   (coordinate measuring machine)
  guidances perpendicular to each
  other and calculation of the
  reference line

- profile template   (rotating profile template device)
  (for surfaces of revolution only)

18.7.8 Assessment of orientation deviations

18.7.8.1 Definition

The deviations of the workpiece surface or the workpiece line (axis, generator line) from an (almost) geometrical ideal reference element (plane or straight line embodiment) are measured. The reference element is to be aligned according to the datum or datum system (parallel, perpendicular, in the specified angle) The orientation of the datum is determined by the minimum-rock requirement at the datum element of the workpiece (see 18.3.2) or by the adjustment of the workpiece in the datum system (see 3.4).

When the reference element does not intersect the workpiece feature (but eventually touches) the orientation deviation δ_d according to ISO 1101 is the difference between the largest and smallest distances of the workpiece feature from the reference element (Table 18.1). When the reference element does intersect the workpiece feature the orientation deviation δ_d according to ISO 1101 is the sum of the largest distances of the workpiece feature from the reference element on both sides of the reference element (Table 18.1).

The orientation deviation δ_d must not exceed the orientation tolerance t_d: $\delta_d \leq t_d$.

Note that this definition is in accordance with ISO 1101 and ISO 5459. The orientation deviation encloses the flatness deviation of the surface to be measured or the straightness deviation of the axis or line to be measured.

According to the Comecon Standard ST RGW 301-76, when measuring the orientation deviation, the form deviation of the feature to be measured is to be eleminated, for example by using a parallel plane plate or a mandrel contacting the workpiece surface and taking the measurements from this auxiliary element (plate, mandrel).

With the definition of the orientation tolerance in ISO 1101, which includes the form deviation of the toleranced feature it was assumed that normally no larger form deviations are permitted by the function than the orientation tolerance. This enables simple inspection (measurement).

18.7.8.2 Combination of features

There are possible deviations of parallelism, perpendicularity and angularity:

- of a plane surface relative to a plane surface;

- of a plane surface relative to a straight line (axis, generator line);

- of a straight line (axis, generator line) relative to a plane surface;

- of a straight line (axis, generator line) relative to a straight line (axis, generator line).

18.7.8.3 Measurement methods

18.7.8.3.1 General For the reference straight line (embodiment of straight line) see 18.7.1.3 and for the reference plan surface see 18.7.3.

The reference angle (embodiment of angle) may be established by

• high-precision divided circle	(dividing head);
• high-precision straightness guidances, perpendicular to each other	(form measuring instrument) (coordinate measuring machine)

- solid angle   (measuring plate with measuring angle)
                                                   (sine bar rule);

- optical beam   (pentaprism).

With the use of mandrels, the cylindricity deviation of the workpiece hole to be measured is eliminated from the measurement. Therefore this method is an approximation according to ISO 1101 (but a method close to the definition according to ST RGW 301-76; see 18.7).

18.7.8.3.2 Assessment of orientation deviations of straight generator lines or plane surfaces by measuring distances For the measurement method see Figure 18.48.
The orientation deviation δ_d (deviation of parallelism, deviation of perpendicularity) is the difference between the largest distance A_{max} and the smallest distance

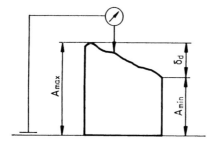

Figure 18.48: Assessment of orientation deviation $\delta_d = A_{max} - A_{min} \leq t_d$ (here the parallelism deviation δ_p) of a surface or generator line.

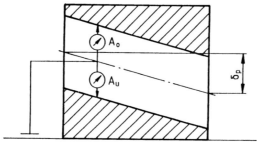

Figure 18.49: Assessment of parallelism deviation δ_p of an axis, evaluated from measurements of distances:

$$\delta_p = \left(\frac{A_o - A_u}{2}\right)_{max} - \left(\frac{A_o - A_u}{2}\right)_{min} \leq t_p.$$

A_{min} (indicator reading) of the workpiece feature from the measuring table or from the solid angle, and must not exceed the orientation tolerance t_d: $\delta_d \le t_d$.

18.7.8.3.3 Assessment of the orientation deviation of an axis by measuring distances For the measurement method see Figure 18.49.

The differences of the distances from the measuring plate are to be measured. During the measurements, the indicators must keep their alignment, but need not be calibrated in height.

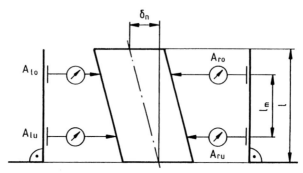

Figure 18.50: Assessment of perpendicularity deviation δ_n of an axis, evaluated from measurements of distances from parallel planes:

$$\delta_n = \frac{(A_{ro} - A_{ru}) - (A_{lo} - A_{lu})}{2} \cdot \frac{l}{l_m} \le t_n.$$

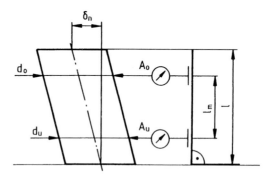

Figure 18.51: Assessment of perpendicularity deviation δ_n of an axis, evaluated from measurements of distances and diameters:

$$\delta_n = \left[(A_o - A_u) - \frac{d_u - d_o}{2} \right] \frac{l}{l_m} \le t_n.$$

As the measurement directions are opposite, the difference of half of the sums of the distances from the measuring plate is assessed.

Figure 18.50 shows the assessment of the orientation deviation with two parallel measuring plates. Figure 18.51 shows the assessment by distance and diameter measurements. A prerequisite of the methods shown in Figures 18.50 and 18.51 is that form deviations are negligible compared with the orientation deviations.

When the tolerance zone is cylindrical, the orientation deviations are to be assessed in two mutually perpendicular axial sections. The cylindrical orientation deviation in this case is

$$\delta_{\mathrm{d}} = \sqrt{(\delta_{\mathrm{d}x}^{2} + \delta_{\mathrm{d}y}^{2})}$$

(see 18.6).

18.7.8.3.4 Assessment of orientation deviations with form measuring instruments or coordinate measuring machines For the assessment of the orientation deviation of a plane surface or a straight generator line see 18.7.8.3.2.

The orientation deviation of an axis can be assessed by measurements of distances as described in 18.7.8.3.3 or by the centres of cross-sections assessed from the circumference lines in cross-sections perpendicular to the axis.

For the assessment of cross-section centres see 18.7.2.1.

18.7.9 Assessment of location deviations

18.7.9.1 Definition

The deviations of the workpiece surface (plane surface) or the workpiece line (axis) from an (almost) ideal reference element (plane or straight line embodiment) are measured. The reference element (embodiment) is to be aligned according to the datum or datum system (parallel, perpendicular, in the specified angle). The orientation of the datum is determined by the minimum-rock requirement at the datum element of the workpiece (see 18.3.4) or by the adjustment of the workpiece in the datum system (see 3.4). The distance between the reference element and the datum (theoretical exact location of the embodiment) is zero (coaxiality, symmetry) or specified by theoretical exact dimensions (in rectangular frames). The same applies to the distances of the reference elements between each other, if applicable.

The embodiment can be located apart from the theoretical exact location, but then the measured values have to be corrected accordingly.

The location deviation δ_{o} according to ISO 1101 is the largest distance of the workpiece feature (surface, axis) from the reference element (which is or is considered to be in the theoretical exact location and orientation) (Figures 18.52 and 18.53) (Table 18.1), and must not exceed half of the locational tolerance t_{o}: $\delta_{\mathrm{o}} \leq t_{\mathrm{o}}/2$.

Note that this definition is in accordance with ISO 1101 and ISO 5459. The

239

location deviation encloses the flatness deviation of the surface to be measured or the straightness deviation of the axis to be measured.

According to the Comecon Standard ST RGW 301-76, when measuring the location deviation, the form deviation of the feature to be measured is to be eliminated, for example by using a parallel plane plate or a mandrel contacting the workpiece surfaces and taking the measurements from this auxiliary element (plate, mandrel) (Figure 18.14).

With the definition of the location tolerance in ISO 1101, which includes the form deviations of the toleranced feature, it was assumed that normally no larger form deviations are permitted by the function than the location tolerance. This enables simple inspection (measurement).

18.7.9.2 Combination of features

There are possible deviations of position:

- of a point relative to a plane surface;
- of a point relative to a straight line (axis, generator line);
- of a plane surface relative to a plane surface;
- of a plane surface relative to a straight line (axis, generator line);
- of a plane surface relative to a plane surface;
- of a plane surface relative to a straight line (axis, generator line);
- of a straight line (axis, generator line) relative to a straight line (axis, generator line);
- of a surface relative to a datum system;
- of a line relative to a datum system;
- of a point relative to a datum system.

There are possible deviations of coaxiality:

- of a straight line (axis) relative to a straight line (axis);
- of a centre point relative to a straight line (axis).

There are possible deviations of symmetry:

- of a straight line (axis) relative to a plane surface (symmetry plane);
- of a plane face (symmetry face) relative to a plane face (symmetry face).

18.7.9.3 Measurement methods

18.7.9.3.1 General For the reference element (embodiment of straight line, of plane) see 18.7.1.3 and 18.7.3. For the embodiment of the orientation see 18.7.8.3.1.

The embodiments are to be located in the theoretical exact distance (zero or specified theoretical exact dimension in rectangular frame) from the datum, or the measured values are to be corrected accordingly.

18.7.9.3.2 Assesment of the location deviation of a straight generator line or a plane surface by measuring distances For the measurement method see Figure 18.52.

The location deviation δ_o is the absolute value of the maximum difference between the measured distance A and the theoretical exact distance A_{th}, and must not exceed half of the location tolerance t_o: $\delta_o \leq t_o/2$.

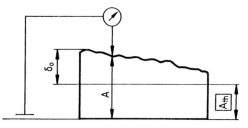

Figure 18.52: Assessment of location deviation of a generator line or a surface: $\delta_o = |A - A_{th}|_{max} \leq t_o/2$.

18.7.9.3.3 Assessment of the location deviation of an axis relative to a plane surface by measuring distances For the measurement method see Figure 18.53.

Both dial indicators are calibrated to the distance from the measuring plate. In each section the arithmetical mean of the distances from the measuring plate $(A_{oi} + A_{ui})/2$ is assessed. The location deviation δ_o is the largest absolute value of the difference between the arithmetical mean and the theoretical exact distance, and must not exceed half of the location tolerance t_o: $\delta_o \leq t_o/2$.

When the tolerance zone is cylindrical, the values $A_i = (A_{oi} + A_{ui})/2 - A_{th}$ (Figure 18.54) must be assessed in each axial location in sections perpendicular to each other, A_{ix} and A_{iy}. The cylindrical location deviation δ_o is the largest value of $\sqrt{(A_{ix}^2 + A_{iy}^2)}$ and must not exceed half of the location tolerance t_o: $\delta_o \leq t_o/2$.

When the theoretical exact distance from the datum is specified by theoretical exact dimensions (rectangular framed) arranged in a chain, the sum of these dimensions applies to the distance of the tolerance zone and reference element from the datum (Figure 18.55).

18.7.9.3.4 Assessment of location deviations with form measuring instruments or

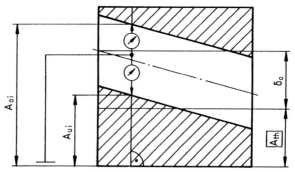

Figure 18.53: Location deviation δ_o of the axis of the hole calculated from distance measurements:

$$\delta_o = \left| \frac{A_{oi} + A_{ui}}{2} - A_{th} \right| \leq \frac{t_o}{2}.$$

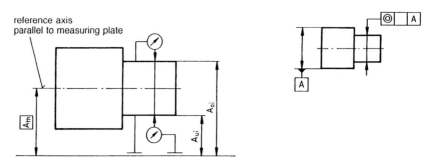

Figure 18.54: Assessment of location deviation δ_o (here coaxiality deviation δ_a) of a shaft by measuring distances:

$$A_i = \frac{A_{oi} + A_{ui}}{2} - A_{th}.$$

coordinate measuring machines For the assessment of the location deviation of a plane surface or a straight generator line see 18.7.9.3.2.

The location deviation of an axis can be assessed by measurements of distances as described in 18.7.9.3.3 or by the centres of cross-sections assessed from the circumference lines in cross-sections perpendicular to the axis.

For the assessment of cross section centres see 18.7.2.1.

18.7.9.3.5 Assessment of coaxiality deviations or symmetry deviations by measuring distances The coaxiality deviation δ_a or the symmetry deviation δ_s are half of

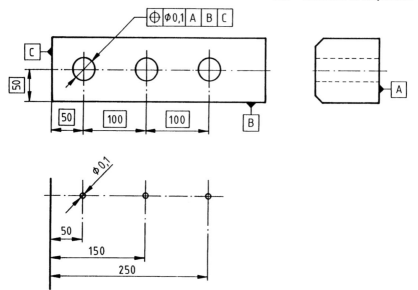

Figure 18.55: Addition of theoretical exact dimensions arranged in a chain.

the largest absolute value of the difference of opposite distances (Figures 18.56 and 18.57), and must not exceed half of the coaxiality tolerance t_a or half of the symmetry tolerance t_s: $|A_{i1} - A_{i2}|_{max}/2 \leq t_a/2$, $|A_{i1} - A_{i2}|_{max}/2 \leq t_s/2$.

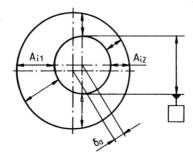

Figure 18.56: Assessment of coaxiality deviation δ_a by measuring distances:

$$\delta_a = \left|\frac{A_{i1} - A_{i2}}{2}\right|_{max}, \quad 2\delta_a = |A_{i1} - A_{i2}|_{max} \leq t_a.$$

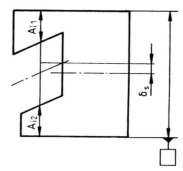

Figure 18.57: Assessment of symmetry deviation δ_s by measuring distances.

$$\delta_s = \left|\frac{A_{i1} - A_{i2}}{2}\right|_{max}, \quad 2\delta_s = |A_{i1} - A_{i2}|_{max} \le t_s$$

18.7.9.3.6 Inspection of coaxiality deviations by measuring run-out deviations Measuring the coaxiality deviation is often complicated and costly. Therefore, for the inspection of the workpiece often the more simple run-out deviations are assessed (see 18.7.10.2.1) and compared with the value of the coaxiality tolerance. Only when the run-out deviation exceeds the coaxiality tolerance is it checked whether twice of the coaxialiy deviation exceeds the coaxiality tolerance (or whether the larger run-out deviation is caused by roundness deviations only, which is permissible)

18.7.9.3.7 Assessment of coaxiality deviations or symmetry deviations with form measuring instruments or coordinate measuring machines The coaxiality deviations or the symmetry deviations can be assessed by measurements of opposite distances as described in 18.7.9.3.5.

The coaxiality deviation can also be assessed by the centres of cross-sections assessed by the circumference lines in cross-sections perpendicular to the axis. For the assessment of cross-section centres see 18.7.2.1.

For the assessment of symmetry faces* the necessary definitions are not yet standardized.

---

*Similar to the way in which the actual axis deviates from a straight line according to the form deviations of the cylindrical surface, the actual symmetry face deviates from a plane according to the form deviations of the plane surfaces establishing the feature. The necessary definitions should be similar to those in 18.7.2.1, but dealing with planes instead of cylinders. They are not yet standardized. See 3.5.

18.7.10   Assessment of run-out deviations

18.7.10.1   Datums

For the assessment of the run-out deviation the workpiece is to be aligned according to the datum axis, or the coordinates are to be transformed accordingly by calculation.

For the datum the minimum-rock requirement according to ISO 5459 applies (see 18.3.4).

The following practical solutions are available.

(a) Mandrel   The mandrel should fit into the hole without clearance. If the mandrel rocks, the minimum-rock requirement applies.

(b) Centres   This is according to the definition when the drawing indicates the centre holes as datum. When the drawing indicates the axis of the cylindrical feature (not the centre holes) as the datum, the measurement is made incorrect by the form deviations and the eccentricity of the centre holes relative to the correct datum axis (of the cylindrical feature).

(c) Chuck   The chuck must have a small run-out deviation in comparison with the run-out tolerance. The suitability can be checked prior to the measurement by measuring the run-out deviation of an almost perfect cylinder embodiment in the chuck. If necessary and possible, the workpiece datum feature is to be aligned within the chuck with the aid of an indicator to indicate the least possible indication difference.

(d) Revolving table   The datum feature is to be aligned with the aid of an indicator in two cross-sections (e.g. one-eighth of the datum feature length apart from the ends) according to the least possible indication difference. This alignment deviates from the theoretical exact datum according to ISO 5459 when the datum feature axis deviates from straightness.

(e) Coordinate measuring machine   From the assessed (probed) points of the datum feature surface the contacting cylinder (see 18.7.2.1) must be determined. When the contacting cylinder can rock, the minimum-rock requirement according to ISO 5459 applies (see 18.3.4). However, in practice the contacting cylinder is aligned parallel to the regression cylinder (because the appropriate mathematical technique has been developed).

(f) Form measuring instrument   Depending on the type of the instrument procedure, (d) or (e) applies.

245

(g) V-blocks, measuring table This method should be applied only when the cylindricity deviation of the datum feature is small in comparison with the run-out tolerance. In other cases the measurement is made considerably inaccurate by the form deviations (see 18.7.4.3).

18.7.10.2 *Measuring methods for circular run-out*

The deviations of the workpiece surface section line from an (almost) geometrical ideal reference circle (circle embodiment) coaxial (concentrical) with the datum axis are measured in sections plane and perpendicular to the datum, or conical or cylindrical and coaxial to the datum. The circle embodiments are established by revolving the workpiece (Figure 18.58) or indicator (Figure 18.59) using

(a) length measuring instrument and workpiece support by

- mandrel between centres,

- centres,

- chuck,

- revolving table;

(b) form measuring instrument.

In coordinate measuring machines the circle embodiment is established by calculation relative to the almost straight and perpendicular axes of the coordinate measuring machine.

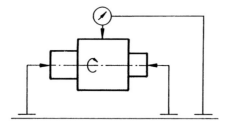 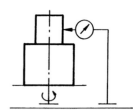

Figure 18.58: Measurement of the run-out deviation with revolving workpiece.

18.7.10.2.1 Assessment of circular radial run-out deviations The deviations of the workpiece circumference line of cylindrical features in sections perpendicular to the datum axis from an (almost) geometrical ideal reference circle (circle embodiment, simulated circle, ISO 5459) are measured. The circle embodiment is coaxial (concentrical) with the datum axis (Figure 18.60).

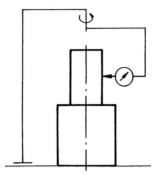

Figure 18.59: Measurement of the run-out deviation with revolving indicator.

When the reference circle does not intersect the workpiece circumference line, the circular radial run-out deviation δ_1 is the difference between the largest, A_{max}, and smallest, A_{min}, distances of the workpiece circumference line from the reference circle, and must not exceed the run-out tolerance t_1. Using dial indicators, this is the difference between the largest and smallest indications during one revolution:

$$\delta_1 = A_{max} - A_{min} \leq t_1.$$

Each section is to be considered independently from the others.

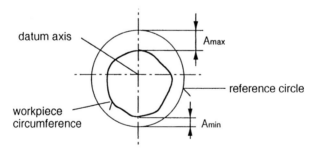

Figure 18.60: Assessment of circular radial run-out deviation $\delta_1 = A_{max} - A_{min} \leq t_1$.

The radial run-out tolerance and deviation are only applicable to cylindrical or sectors of cylindrical features relative to cylindrical datum features.

Note that the circular radial run-out deviation comprises the eccentricity and parts of the roundness deviation of the feature to be measured. When (theoretically) the roundness deviation is zero, the circular radial run-out deviation is twice the eccentricity (Figure 18.61).

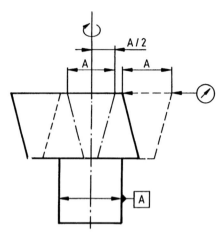

Figure 18.61: Assessment of circular radial run-out deviation $\delta_1 = A \leq t_1$ and eccentricity $= A/2$.

When the workpiece feature to be measured is not very short (so that one measurement is not sufficient), the circular radial run-out deviation should be measured in the centre and near the ends of the workpiece feature.

18.7.10.2.2 Assessment of circular axial run-out deviations The deviations of (circular) section lines of the (plane) workpiece surface in cylindrical sections coaxial with the datum axis (measuring cylinders) from an (almost) geometrical ideal reference plane (with the datum concentric circle in the reference plane, circle embodiment) is measured. The circle embodiment is coaxial (concentric) with the datum axis (Figure 18.62).

When the reference plane does not intersect the workpiece section line, the circular axial run-out deviation δ_1 is the difference between the largest, A_{max} and smallest, A_{min} distances of the workpiece section line from the reference plane (reference circle within the reference plane concentric to the datum), and must not exceed the run-out tolerance t_1. Using dial indicators, this is the difference between the largest and smallest indications during one revolution:

$$\delta_1 = A_{max} - A_{min} = t_1.$$

Each cylindrical section is to be considered independently from the others.

Note that the circular axial run-out deviation is equal to the perpendicularity deviation of the circular section line of the surface but may be smaller than the perpendicularity deviation of the entire surface (Figure 3.8).

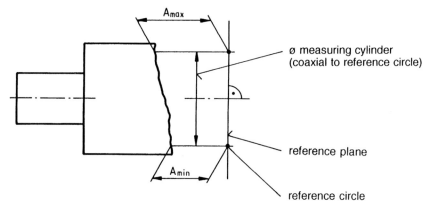

ø measuring cylinder
(coaxial to reference circle)

reference plane

reference circle

Figure 18.62: Assessment of circular axial run-out deviation $\delta_1 = A_{max} - A_{min} \leq t_1$.

The circular axial run-out deviation should be measured at ≈ 1, 0.75 and 0.5 times the outer diameter. With surfaces manufactured by metal removal, the measurement near the outer diameter is in general sufficient, because here the largest run-out deviations occur.

During measurement of the circular axial run-out deviation, the workpiece and indicator must be fixed in the axial direction by using

• a chuck;

• an axial support against an auxiliary datum surface that is plane, perpendicular to the axis and not less in diameter than the measured surface;

• an axial support coaxial to the datum axis;

• a support at the surface to be measured (Figure 18.63).

Support at a surface different from the surface to be measured, where the support is apart from the datum axis, should be avoided because deviations of form and orientation of the surface invalidate the measurement result.

When the support is at the surface to be measured and near the outer diameter, twice the value of the circular axial run-out deviation will be indicated (Figure 18.63).

18.7.10.2.3 Assessment of circular run-out deviations in any or in a specified direction
The deviations of (circular) section lines of the workpiece surface (of rotationally symmetric features) in conical sections coaxial with the datum axis from an

249

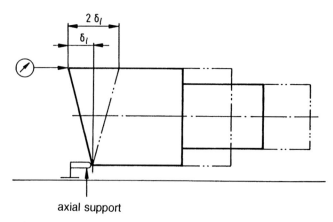

Figure 18.63: Assessment of circular axial run-out deviation δ_1 with axial support at the outer diameter of the surface to be measured.

(almost) geometrical ideal reference circle (circle embodiment) within the conical section and coaxial with the datum are measured. If not otherwise specified, the conical section (measuring cone, measuring direction) is perpendicular to the surface to be measured (Figures 3.25 and 18.64).

When the reference circle does not intersect the workpiece section line, the circular run-out deviation (in any direction or in the specified direction) δ_1 is the difference between the largest, A_{max}, and smallest, A_{min} distance of the workpiece section line from the reference circle, and must not exceed the run-out tolerance t_1. Using dial indicators, this is the difference between the largest and the smallest indications during one revolution:

$$\delta_1 = A_{max} - A_{min} \leq t_1.$$

Each conical section is to be considered independently from the others.

When the workpiece feature to be measured is not very short (so that one measurement is not sufficient), the circular run-out deviation (in any direction or in the specified direction) should be measured in the centre and near the ends of the workpiece feature.

For axial support of the workpiece the same applies as with the circular axial run-out deviation (see 18.7.10.2.2).

Note that the circular run-out deviation in any or the specified direction is composed of the eccentricity and parts of the roundness deviations of the workpiece feature to be measured.

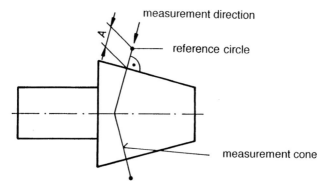

measurement direction

reference circle

measurement cone

Figure 18.64: Assessment of circular run-out deviation δ_l in any direction: $\delta_l = A_{max} - A_{min} \leq t_l$.

18.7.10.3 Assessment of total run-out deviations

The total run-out deviations

• total radial run-out deviation,

• total axial run-out deviation.

are to be distinguished from the circular run-out deviations. With total run-out deviations, the measurement sections are to be considered as dependent on each other. The circle embodiments establish

• with the assessment of the total radial run-out deviation, one reference cylinder co-axial with the datum axis (i.e., using length measuring instruments, the instrument is to be guided along a straight line parallel to the datum axis) (Figure 18.65);

• with the assessment of the total axial run-out deviation, one reference plane perpendicular to the datum axis (i.e., using length measuring instruments, the instrument is to be guided along a straight line perpendicular to the datum axis) (Figure 18.66).

When the reference element (cylinder or plane) does not intersect the workpiece surface, the total run-out deviation δ_t is the difference between the largest, A_{max}, and smallest, A_{min}, distances of the workpiece surface from the reference element, and must not exceed the total run-out tolerance t_t. Using dial indicators, this is the difference between the largest and smallest indications during several (all) revolutions along the workpiece feature:

$$\delta_t = A_{\max} - A_{\min} \leq t_t.$$

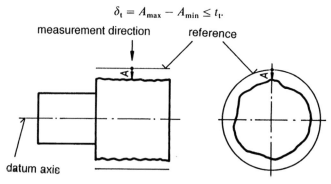

Figure 18.65: Assessment of total radial run-out deviation $\delta_t = A_{\max} - A_{\min} \leq t_t$.

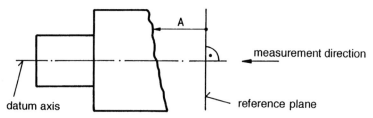

Figure 18.66: Assessment of total axial run-out deviation δ_t.

18.7.11 Inspection of the envelope requirement

18.7.11.1 Definition

It is to be checked whether the surface of the workpiece feature (cylindrical surface or two parallel opposite plane surfaces) is contained within the (almost) geometrical ideal envelope of maximum material size (embodiment, reference element).

18.7.11.2 Inspection methods

The embodiment of the envelope of maximum material size may be established by

- functional gauge (gauge covering the entire length of the feature, e.g. plug, ring, sqared block out of gauge blocks);
- calculation of the rectangular coordinates related to the straight and perpendic-

252

ular guides of the coordinate measuring machine (simulated gauge);

- calculation of the cylinder coordinates related to the straight and perpendicular guides of the form measuring instrument (simulated gauge);

- revolutions on a revolving table, (solid) straight and perpendicular guidance and length measuring instrument;

- revolutions between centres, measuring table and length measuring instrument;

- revolutions in a chuck, measuring table and length measuring instrument;

- revolutions in V-block, measuring table and length measuring instrument;

- measuring table, and length measuring instrument (for features composed of two parallel opposite plane surfaces).

Inspecton with functional gauge   When the gauge covers the entire surface of the workpiece feature, this is an inspection close to the definition (Figure 18.67).

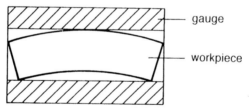

Figure 18.67:   Inspection of the envelope requirement with a functional gauge.

Inspection with coordinate measuring machine   When computer programs are applied to simulate the functional gauge, and sufficient points of the feature's surface are assessed (probed), this is an inspection close to the definition.

Inspection with form measuring instrument   When computer programs are applied to simulate the functional gauge, and sufficient sections of the workpiece feature are assessed, this is an inspection close to the definition.

Inspection with length measuring instrument and measuring table, centres or chuck

253

or revolving table The inspection method is shown in Figure 18.68.

The reference cylinder is established by the axis of revolution parallel to the measuring table (e.g. by centres one of which is adjustable) and a length measuring instrument calibrated to a distance (radius) from the axis of revolution of half of the maximum material size (e.g. calibrated with a disc of known diameter).

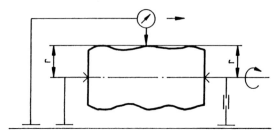

Figure 18.68: Inspection of the envelope requirement: length measuring instrument, measuring table and centres.

When sufficient points of the workpiece feature surface are assessed, and the workpiece is correctly adjusted (e.g. with an adjustable support in order to avoid the effect shown in Figure 18.69), this is an inspection close to the definition.

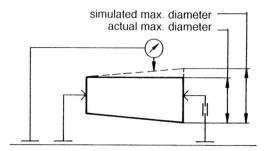

Figure 18.69: Inspection of the envelope requirement: failure caused by insufficient adjustment of the workpiece feature.

Inspection with length measuring instrument, measuring table and/or V-block Revolutions of the workpiece feature in V-blocks or revolutions (rolling) on a measuring table cannot detect certain types of form deviations (oval and lobed), and should be applied only when these are not dominant (see 18.7.4.3).

Inspection with measuring table or measuring plate and length measuring instrument The inspection method is shown in Figures 18.70 and 18.71, which show that it is necessary to inspect from both sides. With outer dimensions (Figure 18.70), the method with the smaller maximum indication applies and with inner dimension (Figure 18.71), the method with the smaller minimum indication.

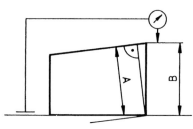

Figure 18.70:   Inspection of the envelope requirement of two parallel opposite plane surfaces on a measuring table with length measuring instrument: $A < B$; smaller value (A) applies.

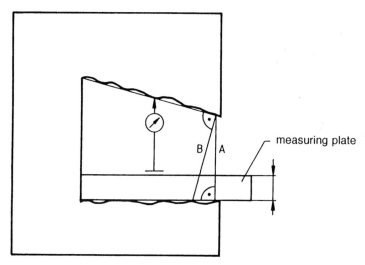

Figure 18.71:   Inspection of the envelope requirement of two parallel opposite plane surfaces with a measuring plate (e.g. gauge block) and a length measuring instrument: $A < B$; smaller value (A) applies.

When sufficient points of the workpiece feature surfaces are assessed, this is an inspection close to the definition.

18.7.12 Inspection of the maximum material requirement

18.7.12.1 Definition

It is to be checked whether the surface of the workpiece feature (cylindrical surface or two parallel opposite plane surfaces) is contained in the (almost) geometrical ideal boundary of maximum material virtual size (embodiment, reference element).

The reference elements at the toleranced and datum features have the (almost) geometrical ideal orientation (parallel, perpendicular, in specified angle) and, when the maximum material requirement is applied to a location tolerance, are located (almost) at the theoretical exact (geometrical ideal) location.

The reference element at the toleranced feature is to be adjusted according to the datum or datum system (parallel, perpendicular, in specified angle). The orientation of the datum is defined by the minimum-rock requirement at the datum feature of the workpiece (see 18.3.4) or by the adjustment of the workpiece in the datum system (see 3.4). The distance between reference element and datum (theoretical exact location of the reference element) is zero (coaxiality, symmetry) or is specified by an theoretical exact (rectangular framed) dimension. The same applies to the distances between the reference elements.

When the maximum material requirement is applied to the datum feature, the reference element at the latter has (almost) geometrical ideal form at maximum material virtual size (see 9.2 and 9.3.3), and replaces the datum.

18.7.12.2 Inspection methods

The reference element (embodiment of the envelope at maximum material virtual size) may be established by

- functional gauge;

- calculation of the rectangular coordinates related to the straight and perpendicular guides of the coordinate measuring machine;

- calculation of the cylinder coordinates related to the straight, perpendicular and circular guides of the form measuring instrument (with coaxiality tolerances to which the maximum material requirement is applied).

As an approximation, the actual sizes of the features and the actual distances of the features can be assessed for the inspection.

18.7.12.2.1 Inspection with functional gauge For sizes and forms of the gauges see 9.2 and 9.3.

When the gauge covers the entire surfaces of the workpiece features this is an inspection close to the definition (Figure 18.72).

a) drawing indications

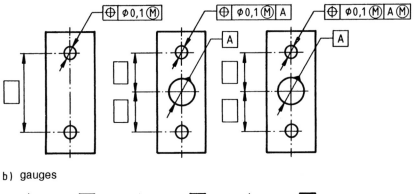

b) gauges

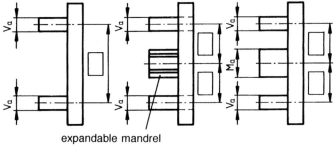

expandable mandrel

Va   max. mat. virt. size
Ma   max. mat. size

Figure 18.72:   Inspection of the maximum material requirement with functional gauges.

18.7.12.2.2 Inspection with form measuring instrument or with coordinate measuring machine   When computer programs are applied to simulate the functional gauge, and sufficient sections or points of the features surfaces are assessed (probed), this is an inspection close to the definition.

18.7.12.2.3 Approximate inspection by measuring sizes and distances   When the deviations of form and of orientation of the features are negligible compared with the location tolerances, the maximum material requirement applied to the location tolerance can be inspected by measuring the actual sizes of the toleranced feature(s), the actual sizes of the datum feature(s) and the actual distance(s) of the features.

The location tolerance may be exceeded by the difference between maximum material size and actual size (when the latter does not take full advantage of the size tolerance).

Positional tolerances without datum   Figure 18.73 shows an example of four holes related to each other by positional tolerances but without datum. Figure 18.74 shows the relevant functional gauge (embodiments). Figure 18.75 shows the tolerance zones for the case when all actual sizes are at the maximum material size ($\varnothing$ 8.1) (and the holes almost of geometric ideal form and orientation related to each other). Figure 18.76 is the dynamic tolerance diagram. It shows the permissible deviation of the actual axes from the geometric ideal location depending on the (for all holes equal) actual sizes. The same information is given in Table 18.5.

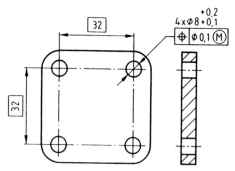

Figure 18.73:   Example of drawing indication, positional tolerancing without datum.

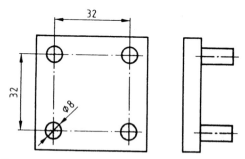

Figure 18.74:   Functional gauge according to the example in Figure 18.73.

Measurements and graphical evaluations are as follows:

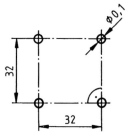

Figure 18.75: Tolerance zones for the maximum material condition according to the example in Figure 18.73.

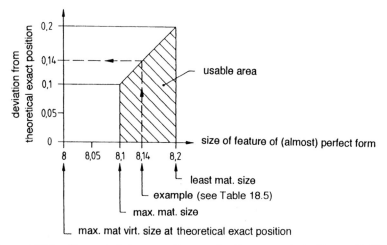

Figure 18.76: Dynamic tolerance diagram for the example in Figure 18.73.

Table 18.5: Positional tolerances for the example in Figure 18.73

Diameter	Positional tolerance
8.1	0.1
8.12	0.12
8.14	0.14
8.16	0.16
8.18	0.18
8.2	0.2

- specify the coordinate system for the measurement, e.g. the actual axis (centre) of the hole left below as coordinate origin, and the actual axis (centre) of the hole right below to determine the x axis (Figure 18.77);

- assess the positional deviations of the holes in this coordinate system;

- plot the positional deviations in an enlarged scale (Figure 18.78);

- use a transparent sheet (templet) with the tolerance zone in the same enlarged scale and move it around until, if possible, all points, are enclosed (Figure 18.78);

- if a point remains outside the tolerance zone, measure the actual size of the feature (hole), enlarge the tolerance zone concentrically by the difference between maximum material size and actual size, and check whether the point is contained within this zone while the other points are contained in the (concentric) original tolerance zone (Figure 18.78).

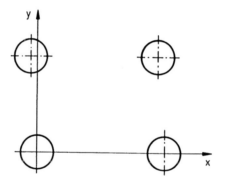

Figure 18.77: Co-ordinate system for the measurement of the workpiece according to Figure 18.73.

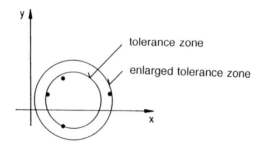

Figure 18.78: Measured points (positional deviations) and tolerance templet of the workpiece according to Figure 18.73.

Positional tolerances with datum When the maximum material requirement applies to the datum, the pattern of tolerance zones (as a whole) may deviate from the theoretical exact location (relative to the datum axis) by the difference between the maximum material size and the actual size of the datum feature (this does not influence the location of the tolerance zones relative to each other).

Figure 18.79 shows an example of four holes related to each other by positional tolerances and related to a datum hole. The maximum material requirement applies to all five holes. Figure 18.80 shows the relevant functional gauge (embodiment). Figure 18.81 shows the location and magnitude of the tolerance zones when

- the toleranced holes   are at maximum material size,
                                                   are at least material size;

- the datum hole   is at maximum material size,
                                                     is at least material size.

Table 18.6 gives the positional tolerances of the toleranced features (four holes), depending on their actual sizes, and the floating zone of the datum feature, depending on its actual size.

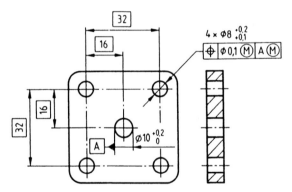

Figure 18.79: Example of drawing indications, positional tolerancing with datum.

Measurements and graphical evaluations are as follows:

- specify the coordinate system for the measurement, e.g. the actual axis (centre) of the datum feature as coordinate origin, and the connection of the actual axes (centres) of the two holes below as the direction of the x axis (Figure 18.82);

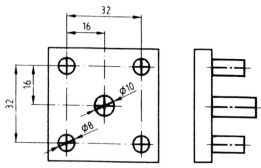

Figure 18.80: Functional gauge according to the example in Figure 18.79.

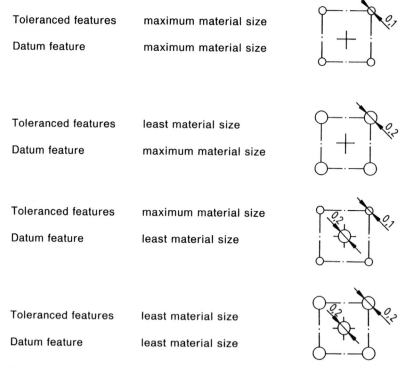

| Toleranced features | maximum material size |
| Datum feature | maximum material size |

| Toleranced features | least material size |
| Datum feature | maximum material size |

| Toleranced features | maximum material size |
| Datum feature | least material size |

| Toleranced features | least material size |
| Datum feature | least material size |

Figure 18.81: Tolerance zones for the maximum material condition and for the least material condition of the example in Figure 18.79.

Table 18.6: Positional tolerances* for the example in Figure 18.79

Diameter of toleranced hole	Positional tolerance of all holes	Diameter of datum hole	Floating zone of datum
8.1 MMS	0.1	10 MMS	0
8.12	0.12	10.05	0.05
8.14	0.14	10.1	0.1
8.16	0.16	10.15	0.15
8.18	0.18	10.2 LMS	0.2
8.2 LMS	0.2		

*Each combination of the values in the second and fourth columns is possible. The values in these columns cannot be added, because they have different effects. Some extreme conbinations are shown in Figure 18.81.

- assess the positional deviations of the holes in this coordinate system;

- plot the positional deviations in an enlarged scale (Figure 18.83);

- use a transparent sheet (templet) in the same scale with concentric
  — tolerance zone,
  — floating zone (difference between maximum material size and actual size of the datum feature,
  and move it around until, if possible, all points are enclosed in the tolerance zone and the coordinate origin is still within the floating zone (Figure 18.83);

- if a point remains outside the tolerance zone, measure the actual size of the feature (hole), enlarge the tolerance zone concentrically by the difference between maximum material size and actual size, check whether the point is contained within this zone while the other points are contained in the (concentric) original tolerance zone and the coordinate origin in the floating zone (Figure 18.83).

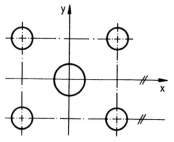

Figure 18.82: Coordinate system for the measurment of a workpiece according to Figure 18.79.

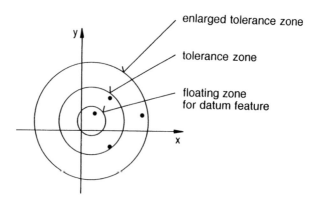

Figure 18.83: Positional deviations and tolerance templet for a workpiece according to Figure 18.79.

Positional tolerances with and without datum Figure 18.84 shows an example of four holes related to each other by rather small positional tolerances and related to the datum features A and B (datum system) by rather large positional tolerances.

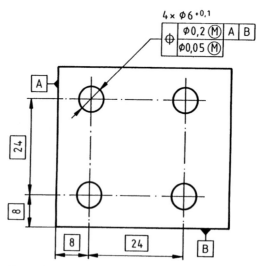

Figure 18.84: Example of drawing indication; positional tolerances with and without datum.

Figure 18.85 shows the relevant functional gauge (embodiment).

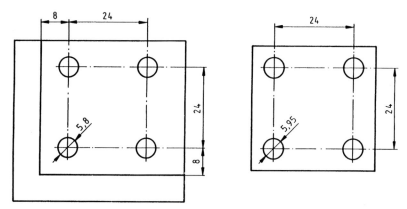

Figure 18.85: Functional gauge according to the example in Figure 18.84.

Figure 18.86 shows the tolerance zones when all actual sizes are at the maximum material size ($\varnothing$ 6). Dynamic tolerance diagrams or tables containing the positional tolerances depending on the actual sizes of the holes similar to Figure 18.76 and Table 18.5 could also be shown.

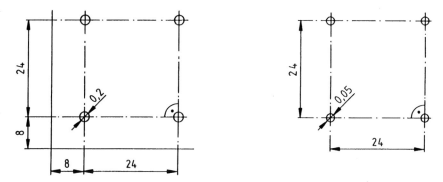

Figure 18.86: Tolerance zones for the maximum material condition of the example in Figure 18.84.

Measurements and graphical evaluations are as follows:

• assess the positional deviations of the holes in the AB coordinate system;

• plot the positional deviations in an enlarged scale (Figure 18.87);

• use a transparent sheet (templet) in the same scale with the tolerance zone $\varnothing$

0.05 (Figure 18.87): the procedure is the same as in 18.7.10.2.3.1;

● use a transparent sheet (templet) in the same scale with the tolerance zone ⌀ 0.2 with centre located at the coordinate origin, check whether all points are contained within this zone: if not, measure the actual size of the concerned hole, enlarge the tolerance zone by the difference between the maximum material size and the actual size, and check whether the point is within this zone (Figure 18.87).

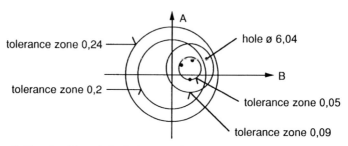

Figure 18.87: Positional deviations and tolerance templets for a workpiece according to Figure 18.84.

18.7.12.2.4 Simplified inspection When no functional gauges are available and coordinate measuring machines or form measuring instruments are not currently used and the prerequisites according to 18.7.12.2.3 do not apply, the workpieces may be inspected in the first step as if the maximum material requirement were not applied. Only when with this inspection the location tolerance is exceeded should it be checked in the second step by a suitable method (e.g. by an coordinate measuring machine), whether or not the maximum material requirement is violated.

A prerequisite for this procedure is that the total tolerance is split into a size tolerance and a location tolerance (and is not indicated as size tolerance together with a zero location tolerance).

18.7.13 Inspection of the least material requirement

18.7.13.1 Definition

It is to be checked whether the surface of the workpiece feature (cylindrical surface or two parallel opposite plane surfaces) violates the (almost) geometrical ideal boundary of least material virtual size (embodiment, reference element).

The reference elements of the toleranced and datum features have the (almost) geometrical ideal orientation (parallel, perpendicular, in specified angle) and, when the least material requirement is applied to a location tolerance, are located (almost) at the theoretical exact (geometrical ideal) location.

The reference element at the toleranced feature is to be adjusted according to the datum or datum system (parallel, perpendicular, in specified angle). The orientation of the datum is defined by the minimum-rock requirement at the datum feature of the workpiece (see 18.3.4) or by the adjustment of the workpiece in the datum system (see 3.4). The distance between reference element and datum (theoretical exact location of the reference element) is zero (coaxiality, symmetry), or is specified by an theoretical exact (rectangular framed) dimension. The same applies to the distances between the reference elements.

When the maximum material requirement is applied to the datum feature, the reference element at the datum feature has (almost) geometrical ideal form at maximum material virtual size (see 9.2 and 9.3.3), and replaces the datum.

When the least material requirement is applied to the datum feature, the reference element at the datum feature has (almost) geometrical ideal form at least material virtual size (see 11.3). The surface of the datum feature must not violate this boundary. The minimum-rock requirement does not apply to this datum (Figure 18.88).

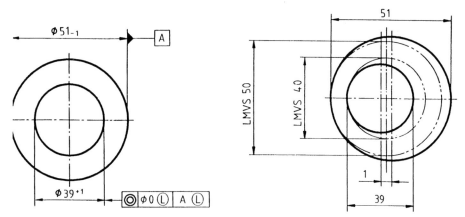

Figure 18.88: Least material requirement applied to the toleranced feature and to the datum feature.

18.7.13.2 Inspection methods

The reference element (embodiment of the boundary of least material virtual size) may be established by

● calculation of the rectangular coordinates related to the straight and perpendicular guides of the coordinate measuring machine;

- calculation of the cylinder coordinate related to the straight, perpendicular and circular guides of the form measuring instrument (with coaxiality tolerances to which the least material requirement is applied).

As an approximation the actual sizes of the features and the actual distances of the features can be assessed for the inspection.

As it must be checked whether the reference element is entirely within the material of the workpiece feature (whether the workpiece feature's surface violates the reference element), gauging is not possible.

Inspection with coordinate measuring machine or with form measuring instrument When computer programs are applied to simulate the geometrical ideal boundary of least material virtual size, and sufficient points or sections of the feature's surfaces are assessed, this is an inspection close to the definition.

Approximate inspection by measuring sizes and distances When the deviations of form and orientation of the features are negligible compared with the location tolerances, the least material requirement applied to the location tolerance can be inspected by measuring the actual sizes of the toleranced feature(s), the actual sizes of the datum feature(s) and the actual distances of the features.

The location tolerance may be exceeded by the difference between the least material size and the actual size (when the latter does not take full advantage of the size tolerance).

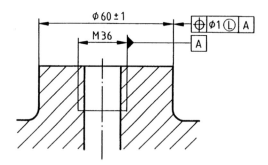

Figure 18.89: Example of drawing indication; positional tolerancing applied to the least material requirement.

Figure 18.89 shows an example of application of the least material requirement. Figure 18.90 shows the relevant geometrical ideal boundary at least material virtual size that is coaxial with the datum A (reference element) (see 11.3).

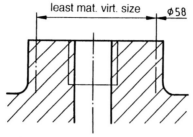

Figure 18.90: Geometrical ideal boundary at least material virtual size for the example in Figure 18.89.

Figure 18.91 shows the positional tolerance zone at least material size and maximum material size. Table 18.7 lists the diameters of the positional tolerance zones of the outer cylindrical feature depending on its actual sizes.

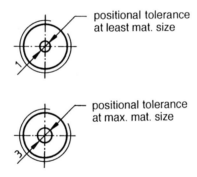

positional tolerance
at least mat. size

positional tolerance
at max. mat. size

Figure 18.91: Positional tolerance zones at maximum material condition and least material condition according to the example in Figure 18.89.

Table 18.7: Positional tolerances for the example in Figure 18.89

Diameter of cylinder	Positional tolerance of cylinder
59 LMS	1
60	2
61 MMS	3

18.7.14 Assessment of the positional deviation for projected tolerance zones

18.7.14.1 Definition

The deviations of a specified extension (projection) of the feature axis from an (almost) geometrical ideal reference straight line (embodiment, simulated straight line, ISO 5459) that is in the (almost) geometrical ideal orientation and location is measured.

The positional deviation δ_c is the largest distance (within the specified length) of the extension of the axis from the reference straight line, and must not exceed half the positional tolerance t_c: $\delta_c \leq t_c/2$.

18.7.14.2 Measurement methods

For the measurement, the extension of the axis to be assessed can be established, for example by

- fitting, as far as possible without clearance, with an (almost) geometrical ideal counterpart (screw, bolt);

- simulation of the fitted counterpart (by calculation) in the coordinate measuring machine.

The positional deviation along the specified extension of the feature axis is measured as described in 18.7.9.3. See also 7.

18.8 ASSESSMENT OF GEOMETRICAL DEVIATIONS OF THREADED FEATURES

If not otherwise specified, the tolerances of orientation, location or run-out of threaded features apply to the pitch diameter (ISO 1101).

The following are used for the measurements:

- form measuring instruments;*

- coordinate measuring machines;*

- special devices (see e.g. Figures 18.92–18.95).

When measuring deviations of orientation, location or run-out for the support in the thread with or without clearance (when the threaded feature is a datum

---

*If suitable probes and computer programs are available or if measured together with special devices as described in this section.

with or without application of the maximum material requirement) and for probing or tracing of threaded features, special measuring devices are to be applied.

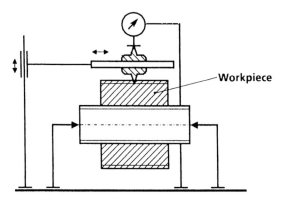

Figure 18.92:   Assessment of coaxiality deviation of a threaded ring.

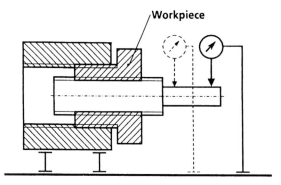

Figure 18.93:   Assessment of coaxiality deviation of a threaded ring.

For the support in the thread without clearance (datum without application of the maximum material requirement), the following may be used:

• conical threaded mandrel (e.g. with a cone angle of 0.5°);

• mandrel or ring with two parts of thread that can be adjusted in the radial or axial direction in order to contact the thread flanks without clearance;

• two thread ring gauges screwed against each other

271

with surfaces to be supported and sufficiently cylindrical and coaxial with the thread.

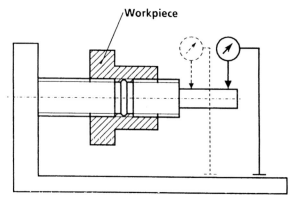

Figure 18.94: Assessment of coaxiality deviation of a threaded ring.

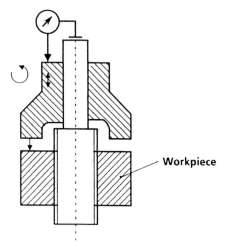

Figure 18.95: Assessement of axial run-out deviation of a nut.

The same devices may be used at the toleranced threaded features to assess the deviations of orientation or location or run-out by measurement at the cylindrical surfaces (see 18.7.8, 18.7.9 and 18.7.10).

For probing at the pitch diameter by a self-centering probing mode, the following may be used:

- ball of diameter suitable for the thread pitch (Figure 18.96 and Table 18.8);

- notch and or measuring cone suitable for the thread pitch (median diameter of the truncated cone = $P/2$; see Figure 18.96);

- thread segments.

When deviations of orientation or location are to be measured with a probing ball by a self-centering probing mode, the probing ball should contact the thread flanks near the pitch diameter. Table 18.8 gives the ball diameters when the contact points are theoretically

- between $0.75H$ and $0.375H$ of the thread profile (see Figure 18.96);

- at the pitch diameter ($0.5H$).

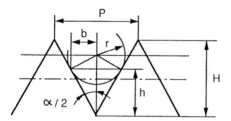

Figure 18.96: Metrical thread, configuration and probing ball radius r for probing near the flanks (using a self-centring probing mode):

$$H = 0.866P, \ \alpha = 60°, \ c = h/H;$$
$$r = \frac{b}{\cos(\alpha/2)} = \frac{h \tan (\alpha/2)}{\cos (\alpha/2)};$$
$$r = cP \times 0.577.$$

D_{rec} in Table 18.8 are probing ball diameters that ensure contact within this range.

When the deviation is measured (probed) directly at the pitch diameter, the stylus must be guided parallel to and symmetrical with the pitch diameter axis, i.e. the probing direction line (axis) must meet the pitch diameter axis. With coordinate measuring machines, this guidance is not necessary and is replaced by appropriate calculations.

When coordinate measuring machines and a small ball probe (smaller than according to Table 18.8), are used the thread flanks are to be assessed in order to calculate the substitute thread flanks (see 8.1). The line of intersection of two adjacent substitute thread flanks is used in order to calculate the pitch diameter

Table 18.8: Recommended probing ball diameters Drec in mm for probing at the pitch diameters of metrical threads according to ISO 261 (P is the pitch in mm)

P	$D_{0.5H}$	$D_{0.375H}$	$D_{0.75H}$	D_{rec}	ISO metric screw threads
0.5	0.29	0.22	0.43	0.3	M3
0.6	0.35	0.26	0.52	0.3	M3,5
0.7	0.40	0.30	0.61	0.5	M4
0.75	0.43	0.32	0.65	0.5	M4,5
0.8	0.46	0.35	0.69	0.5	M5
1	0.58	0.43	0.87	0.5	M6 M7
1.25	0.72	0.54	1.08	0.8	M8 M9
1.5	0.87	0.65	1.30	0.8	M10 M11
1.75	1.01	0.76	1.51	1	M12
2	1.15	0.87	1.73	1	M14 M16
2.5	1.44	1.08	2.16	1.5	M18 M20 M22
3	1.73	1.30	2.60	1.5	M24 M27
3.5	2.02	1.52	3.03	2	M30 M33
4	2.31	1.73	3.46	2	M36 M39
4.5	2.60	1.95	3.89	3	M42 M45
5	2.89	2.17	4.33	3	M48 M52
5.5	3.18	2.38	4.76	3	M56 M60
6	3.46	2.60	5.19	3	M64 to M120
8	4.62	3.46	6.92	5	M125 to M180

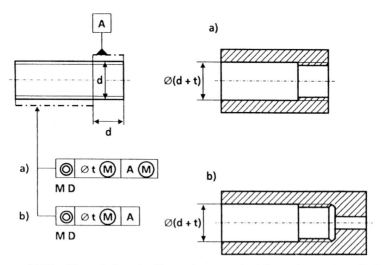

Figure 18.97: Threaded part with maximum material requirement applied to achieve suitability for a stud function (projected tolerance zone, Figure 7.7) drawing indications and gauges.

line according to the theoretical thread configuration (Figure 18.96).

For the inspection of threaded features to which the maximum material requirement is applied, gauges with threads (almost) at the maximum material size are to be applied (see 9 and Figure 18.97), or when coordinate measuring machines are applied and appropriate software is available, the gauges may be simulated.

18.9   TRACING AND PROBING STRATEGIES

18.9.1   General

The workpiece features can be inspected by

- gauging, e.g. with a functional gauge (covering the entire feature);

- areal contacting, e.g. with a measuring plate or mandrel;

- tracing, scanning (continuously, consecutive), e.g. with a dial gauge or form measuring instrument or coordinate measuring machine;

- probing (assessment of a number of points) (discontinuously), e.g. with a coordinate measuring machine.

With tracing, scanning and probing, the surface is assessed by sampling at selected sections or points. Therefore these methods are approximate. The more sections or points are assessed, the more realistically are the geometric deviations assessed and the less is the measuring uncertainty.

Figure 18.98 shows form deviations whose amplitude cannot be fully assessed because the spacing of the probing (assessed points) is too wide.

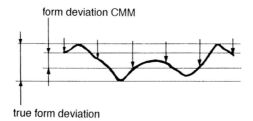

Figure 18.98:   Incomplete assessment of form deviations because of too wide spacing of assessed points.

In order to assess a certain bandwidth of deviations the Nyquist theorem should be respected in all directions of the surface, i.e. the spacing of the assessed points including the distances of sections to be traced or scanned should be at least not more than one half (better not more than one-fifth) of the deviation wavelength to be assessed.

Figure 18.99 shows the effect of the spacing c of the assessed points on a surface exhibiting sinsoidal form deviations of wavelength λ. When the spacing c is greater than half of the wavelength λ the wavelength of the sine wave cannot be assessed (Figure 18.99a). When the spacing is 0.4λ the wavelength of the sine wave will be assessed, but some amplitudes will be considerably reduced (Figure 18.99b). When the spacing c is 0.2λ, the true shape of the sine wave (wavelength and amplitudes) is almost obtained (Figure 18.99c).

With present techniques, this would lead to very time-consuming and costly inspections. Therefore, for economical reasons, the number of selected sections and the number of probed points are normally reduced, but should be distributed in an optimized way.

The necessary numbers of points and sections to be assessed and their optimized distribution depend on

- the type (shape) of the form deviation (depending on the type of manufacturing process);

- the magnitude of the form deviation;

- the ratio of form deviation to geometrical tolerance;

- the geometrical characteristic to be assessed.

Normally for the assessment of, for example, a datum axis or run-out deviations, fewer sections are necessary than for the assessment of cylindricity or total run-out deviations.

International Standards on this subject do not yet exist. However, the British Standard BS 7172 and the former East German Standards TGL 39 093 to TGL 39 098 and TGL 43 041 to 43 045 give some recommendations, which are included in the following. They may serve as a guide in cases where there is no specific information available on the features to be measured that would lead to different strategies (e.g. information on the manufacturing process and on the type and magnitude of the form deviations).

18.9.2 Tracing strategies

Flatness Figure 18.100 shows recommended traces for the assessment of flatness deviations

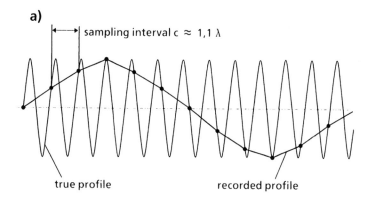

a)

sampling interval c ≈ 1,1 λ

true profile   recorded profile

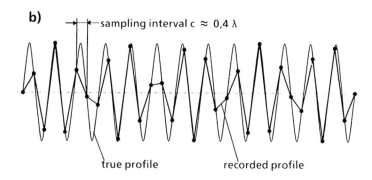

b)

sampling interval c ≈ 0,4 λ

true profile   recorded profile

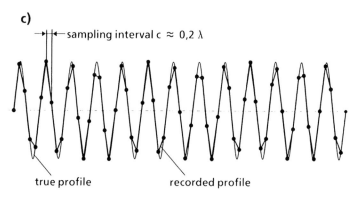

c)

sampling interval c ≈ 0,2 λ

true profile   recorded profile

Figure 18.99: Assessment of sinosoidal form deviations depending on the spacing c of assessed points relative to the wavelength λ of the sine wave.

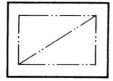

Figure 18.100: Traces for the assessment of flatness deviations.

Roundness Table 18.9 gives the recommended number and location of radial sections (measurement sections) for the assessment of roundness deviations, according to TGL 39 096.

Table 18.9: Recommended number and location of measurement sections for the assessment of roundness

Length l of surface (mm)	≤ 50			$> 50 \leq 250$			< 250		
Ratio l/d of length/diameter	≤ 1	> 1 ≤ 3	> 3	≤ 1	> 1 ≤ 3	> 3	≤ 1	> 1 ≤ 3	> 3
Number N of sections	1	2	3	2	3	4	3	4	5

The distance between the measuring sections is l/N, and the distance between the first or last measuring section and the end of the feature is $l/2N$.

Cylindricity Table 18.10 gives the recommended number of sections and number of traces for the assessment of cylindricity deviations according to TGL 39 097.

Table 18.10: Minimum number of sections and of traces for the assessment of cylindricity deviations

Measurement strategy	Minimum number of sections	Lines
Radial-section method	3 radial sections	3
Generatrix method	3 axial sections	6
Helical method	2 radial sections + 1 helical line of 2 pitches	3
Extreme positions method	1 axial section + 2 radial sections	4

Datum axis Table 18.11 gives the recommended number and location of radial sections (measurement sections) for the assessment of datum axes according to TGL 43 043.

Table 18.11: Recommended number and location of the measuring sections for the assessment of datum axes

Length L_B of datum cylinder (mm)	≤ 50		$> 50 \leq 250$			> 250		
Ratio L_B/d_B of datum cylinder	≤ 3	>3	≤ 1	>1 ≤ 3	>3	≤ 1	>1 ≤ 3	>3
Number n_B of radial sections	2	3	2	3	4	3	4	5
Face distance of radial sections (mm)	$L_B/8$	$L_B/12$	$L_B/8$	$L_B/12$	$L_B/16$	$L_B/12$	$L_B/16$	$L_B/20$
Distance between radial sections (mm)	$3L_B/4$	$5L_B/12$	$3L_B/4$	$5L_B/12$	$7L_B/24$	$5L_B/12$	$7L_B/24$	$9L_B/40$

18.9.3 Probing strategies

The points should be distributed over the entire surface, but not so that they could follow systematic periodic form deviations. For example, six points equally spaced on the circumference of a cylinder cannot detect three lobed form deviations (Figure 18.101). However, seven points equally spaced on the circumference of a cylinder can assess these form deviations by 79% of their amplitudes.

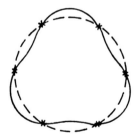

Figure 18.101: Distribution of points with lobed form deviations.

Straight line The length is to be divided into $3N - 2$ subintervals, with points placed in the 1st, 4th, 7th, ... , $(3N - 2)$ subintervals, at random positions (Figure

18.102), where N is the number of points.

Figure 18.102: Distribution of $N = 5$ points on a straight line.

Plane The area is to be divided into $N_1 \times N_2$ rectangles (squares, if possible). Within each rectangle, one point is placed at a random position (Figure 18.103).

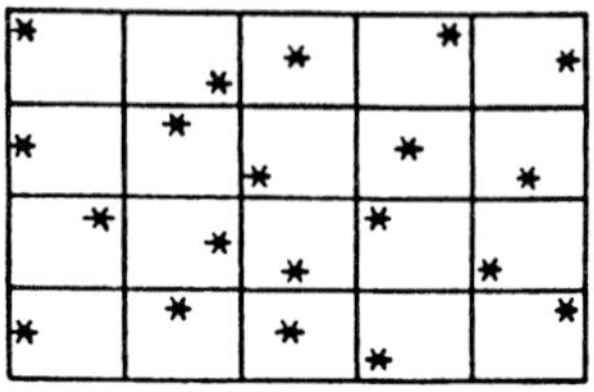

Figure 18.103: Distribution of $N = 20$ points on a plane surface.

When only a small number of points are to be measured, these may be placed in alternate rectangles in a "chessboard" fashion (Figure 18.104).

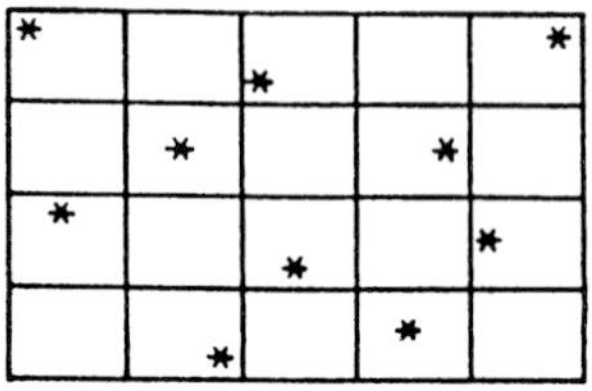

Figure 18.104: "Chessboard" distribution of $N = 10$ points on a plane surface.

Circle N equally spaced points are used, where N, is if possible, a prime number and greater than the expected number of lobes (if a lobed form is to be expected) (Figure 18.101).

Sphere For a sphere sector between two parallel planes, N points are distributed over n_c sections parallel to the end faces and equally spaced with n_p points each (Figure 18.105), where

h = height of sphere section in mm,
r = sphere radius in mm,

n_c = number of sections parallel to the endfaces (including the latter),
n_p = number of points in each section,
N = total number of points,

$n_c \approx \sqrt{(Nh/2\pi r)}$,
$n_p \approx N/n_c$.

For a complete sphere single points at each pole are also to be measured.

Figure 18.105:   Distribution of $N = 30$ points on a spherical surface of $r = 100$ mm and $h = 150$ mm, so that $n_c = 3$ and $n_p = 10$.

Cylinder   (a) Dividing similarly to the plane. (b) Dividing similarly to the sphere.
It is recommended to alternate between odd and even numbers n_p of points on the circles, in order to detect lobings (Figure 18.106).

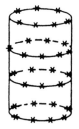

Figure 18.106:   Distribution of 30 points on a cylindrical surface of $r = 10$ mm and $h = 30$ mm ($n_c = 4$, $n_p = 7$ or 8).

Cone N points are used, distributed over n_c sections perpendicular to the axis and equally spaced, where

h = height of truncated cone in mm,

r_1 = radius at the smaller end in mm,
r_2 = radius at the larger end in mm,
k = length of circumference in mm,
n_c = number of sections (endfaces included),
n_p = number of points of a section. (This should decrease towards the vertex
of the cone by the number s; Figure 18.107),

$$k = \sqrt{[h^2 + (r_2 - r_1)^2]},$$
$$n_c \approx \sqrt{[kN/\pi(r_1 + r_2)]},$$
$$s \approx 2\pi(r_2 - r_1)/k.$$

Figure 18.107: Distribution of $N \approx 35$ points over the surface of a truncated cone of $r_1 = 10\,\text{mm}$, $r_2 = 15\,\text{mm}$, $h = 20\,\text{mm}$ ($k = 20.6\,\text{mm}$), so that $n_c = 3$, $s = 2$, $n_p = 10$, 12 and 14.

18.9.4 Number of points

In the absence of form deviations, a minimum number of points would be sufficient to determine a geometric feature in a coordinate system (Table 18.12). Because there are always form deviations, more points have to be assessed. Table 18.12 gives the minimum numbers recommended by BS 7172. Fewer points should not be chosen. More points decrease the measuring uncertainty caused by the form deviations.

In the literature (see Ref. [8]) there is another recommendation to choose at least $8n$ points, where n is the mathematical minimum number of points.

For curved features with changing curvature (e.g. turbine blades) the spacing of the points should be closer for regions of small radii of curvature than for regions of larger radii of curvature.

18.10 SEPARATION FROM ROUGHNESS AND WAVINESS

The separation of roughness and waviness from geometrical deviations is not yet internationally standardized.

Normally the peaks of the surface roughness contribute fully to the geometrical

Table 18.12:   Recommended number of points

Feature	Number of probed points		Remarks
	Mathematical	Recommended	
Straight line	2	5	
Plane	3	9	Distributed on three lines
Circle	3	7	For assessment of three-lobed forms
Sphere	4	9	Distributed on three parallel sections
Cylinder	5	12	For assessment of straightness distributed on four radial sections
		15	For assessment of roundness distributed on three radial sections
Cone	6	12	For assessment of straightness distributed on four radial sections
		15	For assessment of roundness distributed on three radial sections

deviation. How much the surface roughness valleys contribute to the geometrical deviation depends on the measurement method.

In practice the roughness is filtered out by the effect of the stylus tip (ball) of the measuring instrument (with dial indicators, the ball radius is normally 1.5 mm). The effect on the assessment of the geometrical deviation depending on the ball radius and on the spacing of the deviations (irregularities) of the workpiece features surface is shown in Figure 18.108. With surfaces manufactured by metal removal, normally the waviness depth W_t is smaller than $5\,\mu m$ and the waviness spacing wider than 0.3 mm. In these cases the waviness is part of the geometrical deviations. Therefore the corresponding ISO Technical Committees plan to define in an ISO Standard that waviness is part of the geometrical deviations.

Form measuring instruments usually allow the use of a low-pass filter to separate (filter out) roughness (and parts of waviness) in the assessment of geometrical deviations. Figure 18.109 shows the composition of the instruments. Figure 18.110 shows the filter characteristics of these instruments, according to ISO 4291.

The roundness profile of a workpiece can be considered as a superposition of sine waves. Figure 18.111 shows such a Fourier analysis of a roundness profile.

Each sine wave portion will be transmitted by the filter according to the filter characteristic with full or attenuated amplitudes. Waves of short wavelength

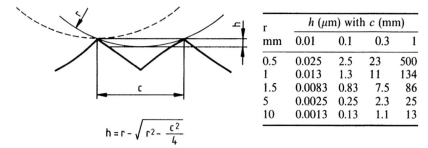

r	h (μm) with c (mm)			
mm	0.01	0.1	0.3	1
0.5	0.025	2.5	23	500
1	0.013	1.3	11	134
1.5	0.0083	0.83	7.5	86
5	0.0025	0.25	2.3	25
10	0.0013	0.13	1.1	13

$$h = r - \sqrt{r^2 - \frac{c^2}{4}}$$

Figure 18.108: Effect of the stylus ball radius and of the workpiece deviation spacings on the assessment of geometrical deviations.

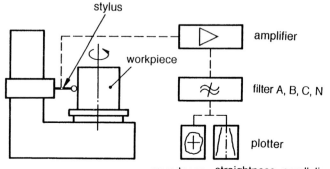

Figure 18.109: Form measuring instrument.

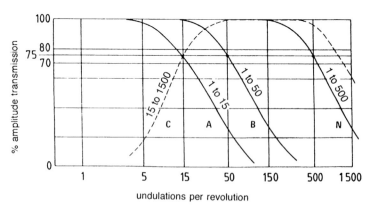

Figure 18.110: Filter characteristics according to ISO 4291.

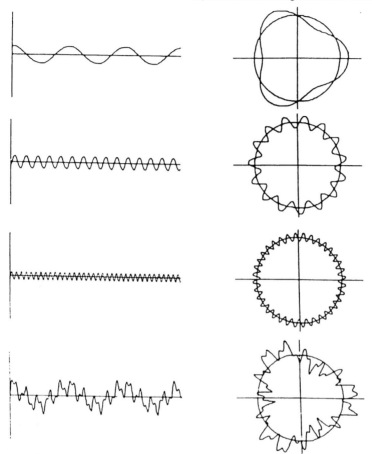

Figure 18.111:   Analysis of a roundness profile according to Wirtz (see Ref. [11]).

(several waves per revolution) (e.g. roughness) will not be transmitted by the filters A and B (or will only be transmitted with strong attenuation).

The filter designations C, A, B and N were used in the past. The filters are now named according to the limiting number of waves (cut-off) n_g or according to the limiting wavelength (cut-off) λ_R. These are the number of sine waves (undulations) per revolution (circumference) UPR or the sine wave length where the amplitudes are transmitted by 75% (old RC-filter) or by 50% (new phase correct Gauss-filter according to ISO 11562).

Figure 18.112 shows an example of the effect of the different filters on the same

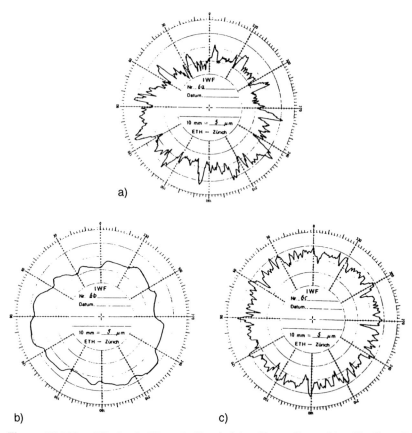

a)

b)   c)

Figure 18.112: Effect of different filters: (a) without filter; (b) with filter B; (c) with filter C and N (according to Wirtz [11]).

workpiece profile to be measured. It shows that the result of the measurement is significantly depending on the filter.

Using narrow-bandpass filters (combination of low-pass and high-pass filters). deviations of certain wavelengths can be selected. Thereby eventually the reasons for certain form deviations (in Figure 18.113 the three-lobed form indicates the effect of the three-point chuck) and the reasons for the functional behaviour become evident. Figure 18.113 shows such an analysis.

In cases of dispute as to whether a workpiece or not complies with the geometrical tolerance the rate of included waviness may be decisive. In these cases it is recommended that agreement be reached on the following:

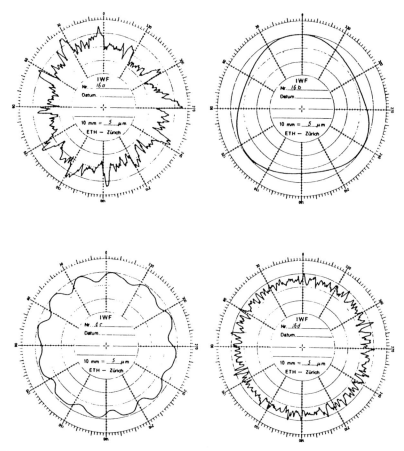

Figure 18.113: Measurement of the same profile with different band passes (according to Wirtz [11]).

- when measuring instruments with filter devices are available, to apply a stylus tip ball radius of 0.5 mm and a filter of 0.8 mm for straight features and according to Table 18.13 for round features;

- when only measuring instruments without filter devices are available, a stylus tip ball radius of 1.5 mm.

This is what will probably be standardized by ISO. If, for example, in international trade other specifications are applied (e.g. according to national standards), this should be specified in the drawing or related documents.

The filters according to Table 18.13 have been calculated so that for a median diameter a filter has a similar effect as a 0.8 mm filter for straight features. The 0.8 mm filter has been choosen because it is the filter that is most frequently

Table 18.13: Filter in UPR for the measurement of roundness and cylindricity

Workpiece diameter (mm)	Filter UPR
>0, ≤8	15
>8, ≤25	50
>25, ≤80	150
>80, ≤250	500
>250	1500

Table 18.14: Low-pass cut-off λ_R for the assessment of straightness and flatness deviations according to TGL 43 041

Roughness R_a (μm)	Roughness R_z (μm)	Cut-off λ_R (mm)
≤0.025	≤0.1	0.25
>0.025, ≤0.4	>0.1, ≤1.6	0.8
>0.4, ≤3.2	>1.6, ≤12.5	2.5
>3.2, ≤12.5	>12.5, ≤50	8
>12.5, ≤100	>50, ≤400	25

Table 18.15: Low-pass filter cut-off n_g for the assessment of roundness, cylindricity, coaxiality, circular radial run-out and total radial run-out deviations according to TGL 39 096

Nominal diameter d (mm)	Cut-off n_g (undulations per revolution, upr) with geometrical tolerance in μm			
	≤2.5	>2.5 ≤5	>5 ≤10	>10
≤10	150	50	50	50
>10, ≤50	500	150	150	50
>50, ≤120	1500	500	500	150
>120, ≤250	1500	1500	500	500
>250	1500	1500	1500	1500

When $n_g = 1500$ upr is not achievable with the measuring instrument, it is permissible to choose the greatest possible cut-oll.

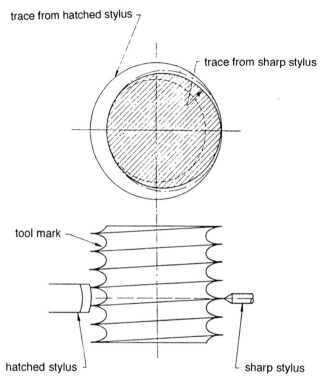

trace from hatched stylus

trace from sharp stylus

tool mark

hatched stylus

sharp stylus

Figure 18.114:   Simulated eccentricity of a turned surface.

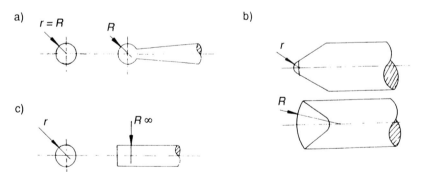

a) $r = R$ R

b) r R

c) r $R \infty$

Figure 18.115:   Stylus tip forms: (a) ball; (b) hatched (toroidal); (c) cylindrical.

used for roughness measurements according to ISO 4288 and for waviness measurements. The tip radius 0.5 mm has been chosen in order to leave its filter effect beyond the 0.8 mm filter effects, i.e. what is filtered out by the tip radius will be filtered out by the 0.8 mm filter anyway, so that the tip radius has no filtering effect on the measurement result.

ANSI B89.3.1 states that, if not otherwise specified, for round features 0.25 mm tip radius and 50 UPR filter apply (see 21.1.6).

In the East European Countries the filter recommended for measurement with form measuring instruments are as shown in Table 18.14 for straightness deviations and in Table 18.15 for roundness, cylindricity, coaxiality and circular radial run-out deviations according to the former East German Standards TGL 43 041 and TGL 39 096, TGL 39 097, TGL 43 042, TGL 43 043. Similar tables for the measurement of axial run-out deviations are given in TGL 43 044.

When the stylus tip (ball) penetrates into the grooves of the roughness (e.g. with turned surfaces), it may occur that the groove is inclined to the section plane of probing so that the stylus tip goes from the peak to the valley within, for example, half a revolution. Thereby an eccentricity is simulated that does not in fact exist (Figure 18.114). This eccentricity remains even when there is a filter involved that separates the roughness from the geometrical deviation. This error will be avoided by the use of a suitable stylus tip form (e.g. a hatched form with radius $r = 10$ mm; Figures 18.108 and 18.115).

18.11 MEASUREMENT UNCERTAINTY

18.11.1 Definition

The measurement uncertainty comprises the random deviations (errors) and the unknown (and therefore uncorrected) systematic deviations (errors) of all quantities that contribute to the measuring result.

It is customary (e.g. according to the German Standard DIN 2257 Part 2) to indicate the measurement uncertainty u with the statistically probability $P = 95\%$.

18.11.2 Direct measurements of geometrical deviations

The measurement uncertainty u_Σ with direct measurement of geometrical deviations is

$$u_\Sigma = \sqrt{(u_1^2 + u_2^2 + \dots + u_n^2)},$$

where $u_1 \dots u_n$ are the uncertainties resulting from influences during the measurement (random or unknown systematic deviations), for example

- deviation of the dimension embodiment;

- deviation caused by the support of the workpiece in the measuring device;

- deviation caused by insufficient sampling (insufficient number of traces or points);

- deviation caused by influences of roughness or waviness of the workpiece;

- deviation caused by temperature;

- deviation caused by measurement force;

- deviation caused by environment, including persons.

The former East German Standards TGL 39 093 to TGL 39 098 and TGL 43 041 to TGL 43 045 indicate further influences, and give hints on measurement uncertainties.

18.11.3   Indirect measurements of geometrical deviations

When geometrical deviations are assessd by indirect measurements, i.e. the deviation δ to be measured is calculated from measured quantities $(x_1, x_2, \ldots, x_i, \ldots, x_n)$ according to the function

$$\delta = F(x_1, x_2, \ldots, x_i, \ldots, x_n),$$

the measurement uncertainty is

$$u_\delta = \sqrt{\left[\sum_{i=1}^{n} \left(\frac{\partial F}{\partial x_i} u_{\Sigma x_i} \right) \right]},$$

where

$\dfrac{\partial E}{\partial x_i}$ = partial derivative of the mentioned function to the measured

   quantity x_i,

$u_{\Sigma x_i}$ = measurement uncertainty of the quantity x_i.

18.11.4   Indication of the measurement uncertainty

According to the rules of quality assurance, the measurement result should comprise the assessed value of measurement corrected by the known systematic deviations (errors) and the measurement uncertainty.

291

Inspection of Geometrical Deviations

According to the rules of measurement, the uncertainty should be less than 10% (at most 20%) of the tolerance to be inspected. In special cases 35% has to be accepted for technical or economical reasons (TGL 39 092).

With the measurement of geometrical deviations, the evaluation of the measurement uncertainty is often complicated, in practice very difficult and time-consuming, and is therefore omitted. An International Standard to this subject, to simplify the procedure, is planned but not yet published.

19 Function-, Manufacturing- and Inspection-Related Geometrical Tolerancing

19.1 DEFINITIONS

19.1.1 Function-related dimensioning and tolerancing

Dimensioning and tolerancing (D&T) are function-related when the functional requirements are directly indicated and toleranced, (Figure 19.1). Then the largest tolerances appear.

Figure 19.2 shows extreme cases of the permissible workpiece shape according to D&T of Figure 19.1. A gauge of width 4 and centre distance 15 must fit into the workpiece.

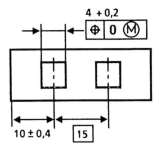

Figure 19.1: Example of function-related D&T (plug fit).

19.1.2 Manufacturing-related dimensioning and tolerancing

D&T are manufacturing related when the dimensions and tolerances that are to be adjusted during manufacturing are directly indicated and when these tolerances can be respected by the intended manufacturing process (Figure 19.3). This is the easiest way to control the manufacturing process. However, the functional requirements, the function-related tolerances, must not be violated. If function-related D&T and manufacturing-related D&T do not coincide, the functional requirements are indicated indirectly. This leads to a reduction in the tolerances.

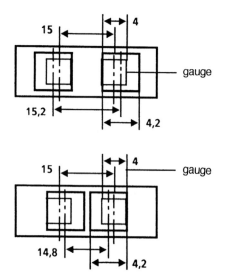

Figure 19.2: Permissible extreme cases according to Figure 19.1.

For example, according to Figure 19.3, the centre distance must not exceed 15.1, but with function-related D&T, according to Figure 19.1, the centre distance may be 15.2 when the actual size of the slot (more precisely, the mating size) is 4.2.

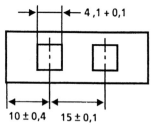

Figure 19.3: Example of manufacturing-related (D&T).

Figures 19.4 and 19.5 demonstrate how the tolerances are derived from the function-related D&T of Figure 19.1 by considering extreme cases. The workpiece must not violate the limits (maximum material virtual condition) indicated in Figure 19.1.

The centre distance may be 14.8 or 15.2 when the slot width is 4.1. Therefore this D&T (Figure 19.3) leads to a reduction in the centre distance tolerance compared with Figure 19.1.

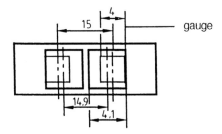

Figure 19.4: Derivation of the least centre distance (14.9) from Figure 19.1.

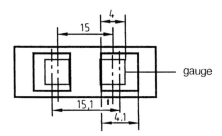

Figure 19.5: Derivation of the maximum centre distance (15.1) from Figure 19.1.

Often, depending on the manufacturing method, there is more than one manufacturing-related D&T. The manufacturing-related D&T according to Figure 19.3 applies to a manufacturing method where the slot centres are to be adjusted. When during manufacture the slot faces are directly adjusted, the manufacturing-related D&T could be according to Figure 19.6.

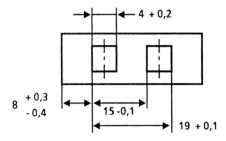

Figure 19.6: Example of manufacturing-related D&T.

Figures 19.7 and 19.8 demonstrate how the tolerances are derived from the function-related D&T of Figure 19.1 considering extreme cases. The workpiece must not violate the limits (maximum material virtual condition) indicated in Figure 19.1.

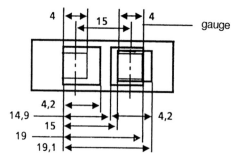

Figure 19.7: Derivation of the limits of size (15 − 0.1 and 19 + 0.1) from Figure 19.1.

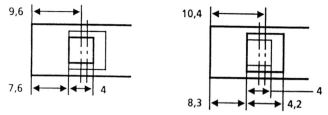

Figure 19.8: Derivation of the limits of size of the edge distance from Figure 19.1.

The distance between the left slot sides may be 14.8 when that between the outer slot sides is 19. However, the distance of 14.8 is not permissible when the distance between the outer slot sides is 19.1. Therefore this D&T (Figure 19.6) leads to a reduction in the tolerances compared with Figure 19.1.

It is also possible to indicate and tolerance the manufacturing process variables (e.g. substitute size and tolerance, form tolerance, substitute distance and substitute location tolerance). Statistical tolerancing may be advantageous. However, appropriate standards are not yet available, see 8 and 14.

19.1.3 Inspection-related dimensioning and tolerancing

D&T is inspection-related when the dimensions and tolerances to be inspected are directly indicated and when the tolerances (deviations) can be inspected with sufficient accuracy. The inspection-related D&T depends on the inspection method. When the maximum material requirement applies and an appropriate gauge is available, the inspection-related D&T coincides with the function-related

D&T. If the inspection uses measurement techniques, the inspection-related D&T may correspond to Figure 19.9. There the total tolerance (Figure 19.1) is split into tolerances of the slot width and of the slot side distances, which leads to a reduction in the slot width tolerances compared with Figure 19.1.

Figures 19.10–19.12 demonstrate how the tolerances are derived from the function related D&T of Figure 19.1 considering extreme cases. The workpiece must not violate the limits (maximum material virtual condition) indicated in Figure 19.1.

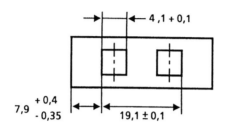

Figure 19.9: Example of inspection-related D&T.

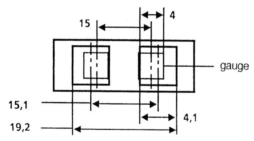

Figure 19.10: Derivation of the minimum inspection size (19.0) from Figure 19.1.

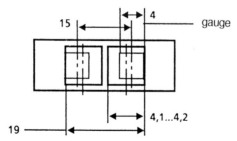

Figure 19.11: Derivation of the maximum inspection size (19.2) from Figure 19.1.

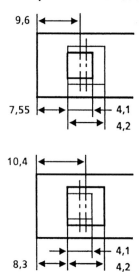

Figure 19.12: Derivation of the inspection sizes of the edge distances from Figure 19.1.

The permissible inspection size may be 19.4 when the slot width is 4.2. However, the inspection size of 19.4 is not permissible when the slot width is 4.1. Therefore this D&T (Figure 19.9) leads to a reduction in the tolerances compared with Figure 19.1.

Indications when inspection is not possible with sufficient accuracy (i.e. when the measurement uncertainty becomes to large) are not inspection-related. For example, tolerancing according to Figure 19.13 (left) is not inspection-related because the datum is too short.

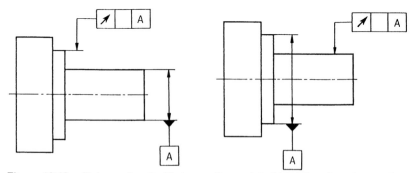

Figure 19.13: Tolerancing that is inspection-related (right) and not inspection-related (left).

If necessary, a further datum is to be indicated (e.g. the flange face) in order to stabilize the orientation of the workpiece (Figure 19.14).

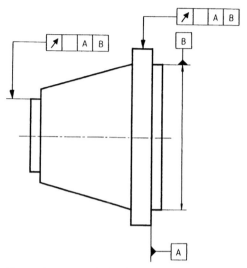

Figure 19.14: Short datum and further datum in order to stabilize the orientation of the workpiece.

19.2   METHOD OF DIMENSIONING AND TOLERANCING

Function-related dimensioning and tolerancing has the advantage of indicating

- the largest tolerances;

- the limits that must not be violated by any combination of variables;

- the limits from which all other methods of D&T are derived.

Because of these advantages, function-related D&T should always be provided. Sometimes (e.g. with mass production) further drawings, manufacturing plans or inspection plans with manufacturing-related or inspection-related D&T derived from function-related D&T can be helpful. CAD provides the possibility of storing the function-related D&T, the manufacturing-related D&T and the inspection-related D&T on different layers of the same drawing data.

Sometimes general guides can be helpful indicating in what functional cases and in what ways manufacturing-related or inspection-related D&T can be derived from function-related D&T. Such guides can, for example, specify that in cases where coaxiality tolerances are indicated but no suitable form measuring instruments or coordinate measuring machines are available, the radial run-out tolerance of the same value shall apply. The same applies in cases where coaxiality tolerances with the maximum material requirement are indicated and no appropriate gauges are available. Then only in cases of dispute need the more precise and more expensive inspection method of the coaxiality tolerance be applied.

Sometimes (e.g. in the single production of heavy machinery) function-related D&T is intentionally avoided (e.g. coaxiality tolerance), and inspection-related D&T (e.g. run-out tolerance) only is used. This when the tolerance is large enough for manufacturing purposes, and expensive inspection methods are to be avoided.

In many functional cases (e.g. with clearance fits) there is a mutual dependence of size and form or orientation or location, which is to be indicated by additional symbols (e.g. Ⓜ; see e.g. Figure 19.1). When these symbols are omitted, the drawing specifies unnecessary restrictions, i.e. there may be parts that violate the drawing indication but that are still suitable for the function. This may lead to uneconomical production. Therefore it should be investigated which of the following is more costly:

- the disadvantages (costs) of these unnecessary restrictions; or

- the greater efforts (costs) in design and in the education in the departments of design, production planning, manufacturing and inspection necessary for handling function-related D&T (providing the largest possible tolerances).

Normally, especially in long series and mass production, function-related D&T (providing the largest possible tolerances) is more economical, although it does require education of staff (see 9.4).

Sometimes application of the symbol Ⓜ can be avoided by the application of positional tolerancing or profile tolerancing of single surfaces (Figure 20.67). However, this has the disadvantage that different theoretical exact dimensions are allocated to part and counterpart, and the functional relationship is more difficult to recognize. Therefore it is recommended that in these cases the symbol Ⓜ be applied (Figure 20.65).

Geometrical tolerances specify geometrical tolerance zones. Particular manufacturing or inspection methods are not specified by geometrical tolerances (however, the tolerances must be respected and verified by the chosen method). Whether inspection is necessary, and to which level and by which method, depend on

- the level of control of the manufacturing process;

- the probability of occurance of deviations (e.g. of lobed forms);

- the consequences of exceeding the tolerance;

● confidence in the manufacturer (e.g. inspection history).

Specification of particular inspection methods would force manufacturers to provide particular inspection devices, in addition to those already possessed for geometrical tolerancing, thus leading to a permanently expansion in instrumentation. Further, there is often a contradiction between the geometrical definition of the geometrical tolerance and the inspection method. Therefore it would be necessary to specify all details of the inspection methods, which would lead to a multiplication of inspection methods and would be detrimental to technical advance. Therefore ISO 1101 does not provide a drawing indication of the inspection method.

19.3   ASSESSMENT OF FUNCTION-RELATED GEOMETRICAL TOLERANCING

The following lists give hints of how to proceed in assessing function-related geometrical tolerancing.

19.3.1   Tolerancing of orientation, location and run-out

● Which features are related to each other by the function?

● Is the general geometrical tolerance sufficient (e.g. ISO 2768 P.2)? If it is not then

● Does one feature determine* orientation or location and may therefore serve as a datum?

● Is it appropriate to specify a common tolerance zone and thereby avoid specification of a datum?

● Is it appropriate to specify a datum system or datum targets?

● Which characteristics should be toleranced?

● Is Ⓜ, Ⓛ, Ⓡ, Ⓟ or Ⓕ appropriate?

● What is the magnitude of the tolerance?

---

*Often (e.g. with clearance fits associated with the maximum material requirement Ⓜ) the features are of the same functional priority with regard to choice as a datum. Then each may be chosen as a datum (Figures 20.47–20.50). However, the dependence shown in Figure 18.9 is to be observed. In contrast, in the case of interference fits, according to Figure 20.95, the assembly direction determines the datum.

19.3.2 Tolerancing of form

● Which characteristic should be toleranced?

● Is Ⓔ appropriate and sufficient? If it is not then

● Is the general geometrical tolerance sufficient (e.g. ISO 2768 P.2)? If it is not then

● Is the form deviation already limited sufficiently by a tolerance of orientation or location or total run-out? If it is not then

● What is the magnitude of the tolerance?

19.3.3 Type of tolerance

● Fits: Ⓔ.

● Clearance fits, but no kinematics: Ⓜ (then often Ⓔ is not needed, see 20.7.3).

● Geometrical ideal form within the material required: Ⓛ.

● Weight limitation: thickness tolerance (maximum thickness).

● Interference fit, kinematics, optics, electrical contacts, measuring contacts: geometrical tolerances regardless of other tolerances.

● Threaded holes, holes for pins: Ⓟ.

● Flexible parts: Ⓕ, ISO 10579-NR.

19.4 ASSESSMENT OF THE OPTIMUM DIMENSIONING AND TOLERANCING

Function-related D&T depends only on the design. However, some designs provide larger and more economical tolerances than others for the same functional purpose. The value analysis provides helps to recognize this.

Whether in addition to the function-related D&T other specifications should be issued providing manufacturing- and/or inspection-related D&T, or whether the drawing should be modified, depends on economic considerations. Participants in these economic considerations should be all concerned departments.

Design department   This is responsible for the tolerances guaranteeing functionality.

Manufacturing planning department   This is responsible for the tolerances being achievable.

Manufacturing department   This is responsible for the tolerances being respected.

Inspection planning department   This is responsible for the tolerances allowing appropriate inspection.

Inspection department   This is responsible for the tolerances being respected.

Standardization department   This is responsible for all concerned receiving the necessary information on standards and their applications.

With the cooperation of these departments, the optimum D&T should be elaborated for typical parts, which afterwards become models for the design department. Experience has shown that without such cooperation no economical D&T can be achieved.

20 Examples of Geometrical Tolerancing

20.1 RESTRICTIONS OF GEOMETRICAL TOLERANCING

20.1.1 Restricted range of application

Figures 20.1 and 20.2 show examples of restricted ranges of application of geometrical tolerances.

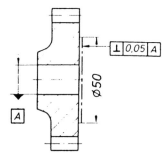

Figure 20.1: Restricted application of perpendicularity tolerance.

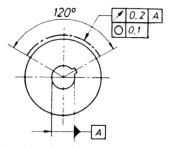

Figure 20.2: Restricted application of run-out and roundness tolerance.

20.1.2 Geometrical tolerances and additional smaller tolerances of their components

It may be that, besides the geometrical tolerances, components of them may have to have narrower tolerances. For example, journal shafts may require smaller

305

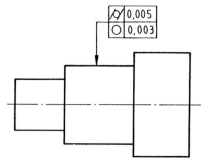

Figure 20.3: Cylindricity tolerance and smaller tolerance of its component roundness.

tolerances for roundness than for cylindricity in order to maintain proper lubrication (Figure 20.3).

According to Figure 20.4, the surface of the right cylinder may vary between two cylinders coaxial with the datum A. According to Figure 20.5, the parallelism deviation (and the straightness deviation) of the generator lines are restricted to a smaller tolerance (0.005).

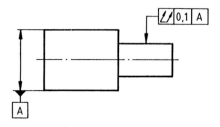

Figure 20.4: Total run-out tolerance.

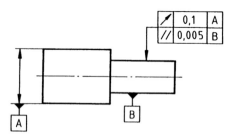

Figure 20.5: Run-out tolerance and smaller tolerance of parallelism of the generator lines.

20.1.3 Superposition of positional tolerances

Figure 20.6 shows two groups of holes. The four hole group is related to the datums D, A and B. The maximum material virtual condition of this group establishes the datum C of the positional tolerances $\varnothing$ 0.15 of the three-hole group. The holes of the latter group must also meet the smaller positional tolerances $\varnothing$ 0.25 relative to each other.

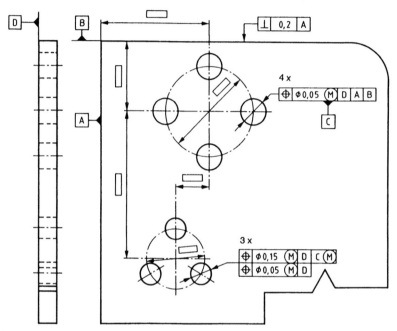

Figure 20.6: Superposition of positional tolerances; group of features (holes) as datum.

20.2 TOLERANCES OF SECTION LINES

When flatness tolerances are indicated, the maximum permissible deviation in straightness is equal in each direction (Figure 20.7a). However, sometimes (e.g. in the case of slideways) the form deviation may be greater in one direction than in the other: Figure 20.7 (b) shows an example.

Parallelism tolerances of surfaces define three-dimensional tolerance zones in which the surface must be contained (Figure 20.8a). When (e.g. for slideways) the parallelism deviation in one direction may be greater than in the other, the parallelism tolerance has to be applied to section lines (Figure 20.8b).

Examples of Geometrical Tolerancing

(a)

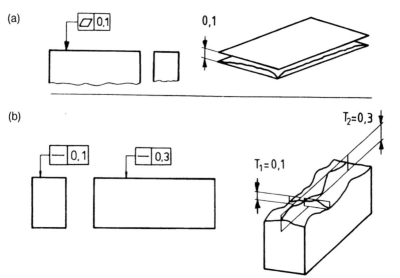

(b)

Figure 20.7: Flatness tolerance (a) and different straightness tolerances (b) of the same surface.

(a)

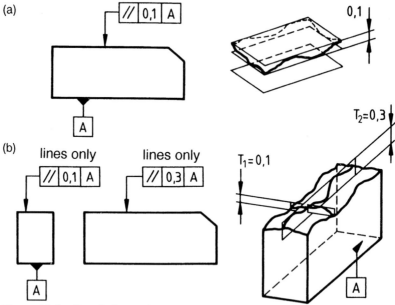

(b)

lines only   lines only

Figure 20.8: Parallelism tolerance of· (a) surfaces; (b) section linos.

308

The indication "lines only" is not necessary when the drawing shows that only lines can possibly be applied (Figure 20.9) (in this case the line of the highest points instead of section lines; see 3.3.1).

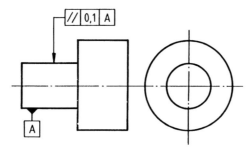

Figure 20.9: Parallelism tolerance of lines of a cylinder.

20.3 TOLERANCES OF PROFILES

With curved profiles (e.g. of turbine blades) the profile tolerance sometimes varies steadily along the curve. Figure 20.10 shows the drawing indication and the relevant tolerance zone. The curvature of the tolerance zone boundary between the indicated points is not standardized.

Figure 20.11 shows a profile tolerance of a curved tube. To limit buckling (e.g. that caused by supports during the bending procedure), the profile tolerance for

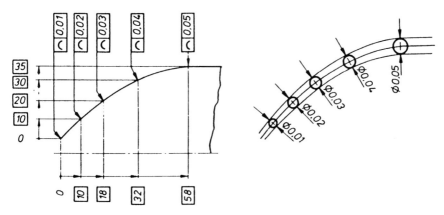

Figure 20.10: Profile tolerance of varying magnitude.

Examples of Geometrical Tolerancing

a limited length and the roundness tolerance are indicated.

Figure 20.12 shows an example of different tolerance zones along a surface. The form of transition from one zone to the next is not standardized, and may require the indication of a note (e.g. "tangential transition").

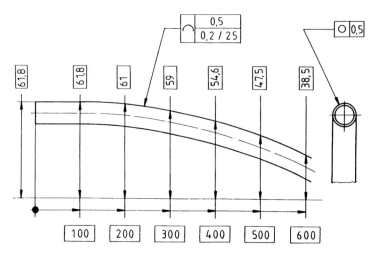

Figure 20.11: Tolerancing to limit buckling.

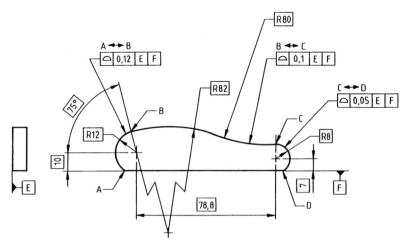

Figure 20.12: Different tolerance zones along a surface.

20.4   POSITION OF A PLANE

The indications (a) and (b) in Figure 20.13 are permissible and define the same tolerance zone. As not only the orientation and form (flatness) but also the location (related to datum B) are important, the indication (c) (orientation tolerance) is not appropriate. ISO 1101 uses the indication (a), ANSI Y14.501 uses the indication (b).

In Figure 20.14 the actual local sizes are limited by the positional tolerance of one surface and the flatness tolerance of the other. The actual local sizes may vary between 49.85 and 50.05.

Figure 20.15 shows the tolerancing of the positions of two inclined plane

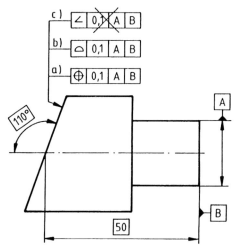

Figure 20.13:   Tolerancing of the position of a plane surface.

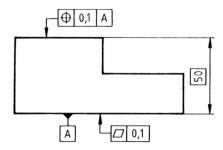

Figure 20.14:   Tolerancing of the position of a plane surface parallel to a datum plane with a flatness tolerance to the datum surface.

surfaces. The theoretical exact position is determined by the datum A (median plane of the slot), the inclination (5°) and the position of the lines (distance 10 from datum A and distance 6 from datum B). The positional tolerances also limit the symmetry deviation of the wedge. The eccentricity (distance of the median face of the wedge from the datum median plane A) can vary up to 0.025.

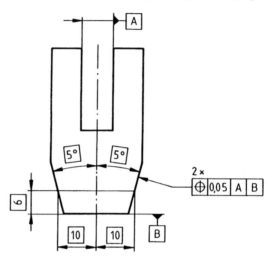

Figure 20.15: Positional tolerancing of two inclined plane surfaces (wedge).

20.5 PERPENDICULARITY TOLERANCES IN DIFFERENT COMBINATIONS

Figure 20.16 (top) shows a perpendicularity tolerance and the permissible workpiece (e.g. a bent workpiece). Figure 20.16 (bottom) shows the tolerancing limiting all perpendicularity deviations.

20.6 LOCATION OF AXES AND MEDIAN FACES

20.6.1 Angular location of features

In Figure 20.17 the keyways are drawn co-symmetrically. The symmetry deviation is limited by the general tolerance.

This limitation has no economical disadvantage, even when the function allows

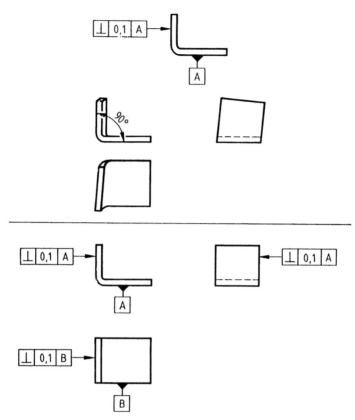

Figure 20.16: Effect of perpendicularity tolerances.

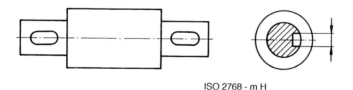

ISO 2768 - m H

Figure 20.17: Co-symmetrical keyways.

a random angular location, because the general tolerance is respected anyway without any additional effort. However, if the angular location is optional, this may be indicated on the drawing (Figure 20.18).

313

Examples of Geometrical Tolerancing

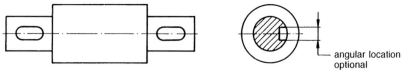

Figure 20.18: Co-symmetrically drawn keyways, but with angular locations optional (also between each other).

20.6.2 Co-symmetry of holes in line

Figure 20.19 shows an example in which the axes of the holes must not deviate more than ±0.05 from a plane. The distances of the axes to the upper plane may vary within ±0.5.

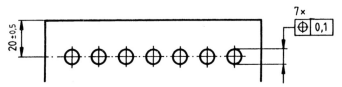

Figure 20.19: Co-symmetry of holes in line.

20.6.3 Crossed axes

Figure 20.20 shows examples of tolerances for the distances of crossed axes.

In case (a) the projection of the horizontal axis at a distance of 75 (i.e. for this point and over the length zero) must not be apart from the projection of the datum axis A for more than 0.025. (The point of the projected axis at distance 75 must be contained between two parallel planes 0.05 apart; their median plane contains the datum axis A.) This tolerance does not limit the angle between the two axis.

In case (b) the projection of the horizontal axis must be contained in a tolerance cylinder of ∅0.05 over a length of 75. The axis of the tolerance cylinder meets the datum axis A; the two axes form an angle of 90°. In this case not only the distance between the workpiece axes is toleranced but also the deviation from the 90° angle between them.

Figure 20.21 shows three different indications (a), (b) and (c) for tolerancing of the perpendicularity deviation of two holes. The indication (a) or (b) alone does not limit the deviation perpendicular to the projection plane (eccentricity). The additional indication (c) limits the eccentricity.

314

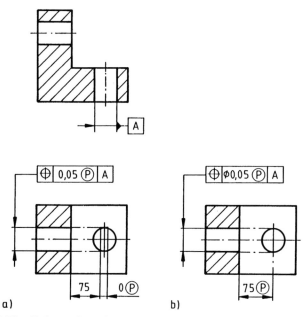

Figure 20.20: Tolerancing of crossed axes.

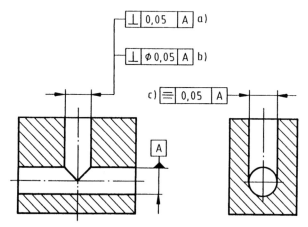

Figure 20.21: Perpendicularity tolerances for axes of holes: (a) and (b) alone do not but the additional indication (c) does limit the eccentricity.

The indication in Figure 20.22 limits the eccentricity of the axes of the holes in relation to the datum axis B. In the case of the six holes the perpendicularity deviation in relation to the datum axis B is also limited.

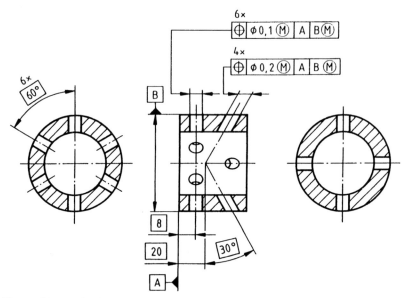

Figure 20.22: Positional tolerances for axes of holes in relation to a datum axis.

20.6.4 Different locational tolerances for different features drawn on the same centre line

When on a drawing features are drawn on the same centre line, it is difficult to

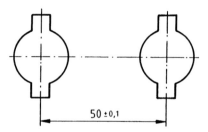

Figure 20.23: Distance tolerance of features drawn on the same centre line.

allocate different distance tolerances to the different features. For example in Figure 20.23 the indication applies for the holes as well as for the slots. Positional tolerancing, however, provides the possibiliy of allocating different tolerances to the different features (Figures 20.24 and 20.25).

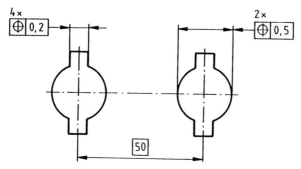

Figure 20.24: Features drawn on the same centre line, but with different positional tolerances.

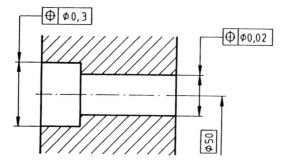

Figure 20.25: Features drawn on the same centre line, but with different positional tolerances.

In Figure 20.26 holes and countersinks have different positional tolerances with different datums. The datum for the positional tolerance of the countersink is the relevant hole axis.

The positional tolerances according to Figure 20.27 allow positional tolerancing of the median faces of the slotted holes in relation to the median plane A. With the indication 10 ± 0.1 instead of the positional tolerances (Figure 20.28), the location of the slotted holes depend on the actual locations of the two cylindrical holes.

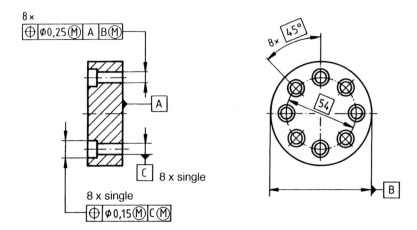

Figure 20.26: Features drawn on the same centre line, but with different positional tolerances related to different datums.

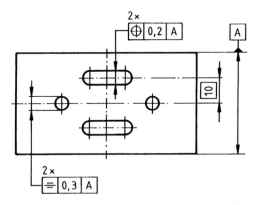

Figure 20.27: Differentiation of locational tolerances in relation to the same median plane.

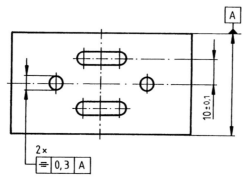

Figure 20.28: Location of the slotted holes dependent on the actual location of the cylindrical holes.

20.6.5 Positional tolerances for features located symmetrically to a symmetry line and multiple patterns of features

ISO 5458 states that positionally toleranced features drawn on the same centre line are regarded as related features having the same theoretical exact location (see 6.2). However, sometimes (e.g. when the features are drawn not on but symetrically with respect to a common centre line), there might be doubt as to what applies. In such cases an appropriate note should be added to the tolerance indication.

Figure 20.29 and 20.30 each show two patterns of holes ($\varnothing$5 and $\varnothing$10). The features of the patterns are drawn on different centre lines but symmetrically with respect to a symmetry line. From this it is not quite obvious whether the zero Ⓜ tolerances apply simultaneously to the two patterns so that they form a single pattern to which a common gauge applies or whether the zero Ⓜ tolerances apply separately to each pattern so that separate gauges apply to the patterns. A similar case applies to the 0.5 Ⓜ tolerances.

In the case of Figure 20.29 the indication "separate requirements" allows both patterns of holes to be regarded as separate patterns.

In the case of Figure 20.30 the indication "simultaneous with ..." ensures that both patterns are to be regarded as a single pattern to which a common gauge applies.

When positionally toleranced features are located symmetrically with respect to a symmetry line (common centre line), and it is not indicated whether the positional tolerances apply separately or simultaneously, it is recommended eitto ask the designer what should apply or to use the failsafe method of simultaneous positional tolerances.

319

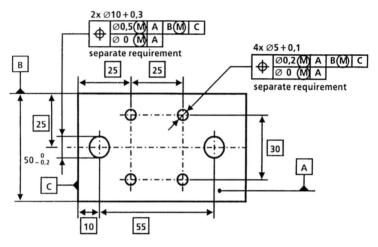

Figure 20.29: Two patterns of holes to be regarded as separate patterns.

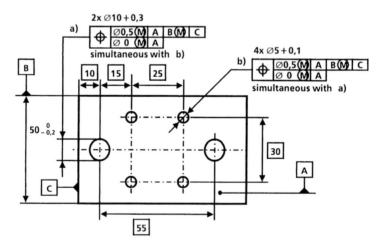

Figure 20.30: Two patterns of holes to be regarded as a single (common) pattern.

ANSI Y14.5M states that multiple patterns of features (two or more patterns of features not necessarily located symmetrically to each other) located by theoretical exact dimensions relative to the same datum system (same datum features referenced in the same order of precedence and in the same modification, i.e. with the same modifier or without modifier) are to be regarded as a single pattern (simultaneous requirements). If this is not required this should be indicated by "SEP REQT" under each tolerance frame (instead of "separate requirement") in Fig. 20.29.

20.7 DATUMS

20.7.1 Plane surface as datum for a symmetry tolerance

Normally the datum for a symmetry tolerance is a median plane or an axis. In special cases it may be a plane surface. Then the simulated datum feature has to be positioned according to the minimum-rock requirement (Figure 20.31).

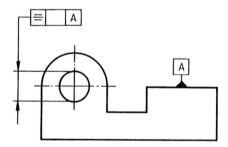

Figure 20.31: Plane surface as datum for a symmetry tolerance.

20.7.2 Plane surface as datum for a run-out tolerance

Normally the datum for a run-out tolerance is an axis. With machines (e.g. electrical motors) it might happen that the datum axis cannot be detected. In such cases the housing base or housing flange may be indicated as the datum (Figure 20.32).

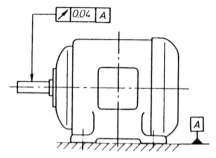

Figure 20.32: Plane surface as datum for a run-out tolerance.

20.7.3 Envelope requirement and maximum material requirement with datums

Figure 20.33 shows on the left a hole to which the envelope requirement Ⓔ applies. On the right the maximum material requirement Ⓜ applies to the datum hole A. This datum hole is a primary datum, and has no form tolerance with a following symbol Ⓜ. This (for the datum hole) has the same effect as the envelope requirement (it must be possible to fit a cylindrical gauge of maximum material

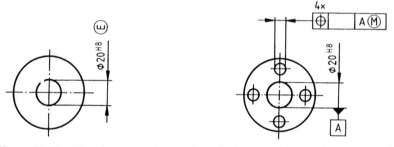

Figure 20.33: Envelope requirement and datum with maximum material requirement.

size). Therefore in tolerancing the hole on the right, the symbol Ⓔ is omitted.

Note that ISO 2692: 1988 contains on p20 (Figure 29a) the symbol Ⓔ behind the size tolerance of the datum, although at the latter the maximum material requirement Ⓜ is applied. The indicaton Ⓜ, however, implies the requirement of Ⓔ. In order to avoid duplication of requirement indications, Ⓔ should be omitted or enclosed in parentheses.

20.7.4 Positional tolerances with and without datums

Positional tolerances may be applied with or without datums. Without a datum, all features combined by the positional tolerance have the same positional tolerance (Figure 20.34).

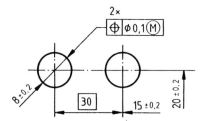

Figure 20.34: Positional tolerances of a group of holes without datum.

When one hole is indicated as a datum (Figure 20.35), this hole has no positional tolerance in relation to the other hole(s). According to Figure 20.35, the gauge has maximum material size = $\varnothing$ 7.8 at the datum and maximum material size minus positional tolerance = $\varnothing$ 7.7 at the other feature. According to Figure 20.34, the gauge has $\varnothing$ 7.8 at both features.

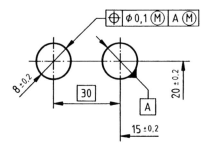

Figure 20.35: Positional tolerances of a group of holes with datum hole.

20.7.5 Sequence and maximum material requirement for datums

Figure 20.36 shows different possibilities for datums.

In case (a) the surface B must contact the gauge according to the minimum-rock requirement. The diameter of the gauge is equal to the maximum material size of the datum feature A.

In case (b) the gauge must enclose the datum feature A with the least possible diameter (mating size of datum feature A). The datum feature is orientated

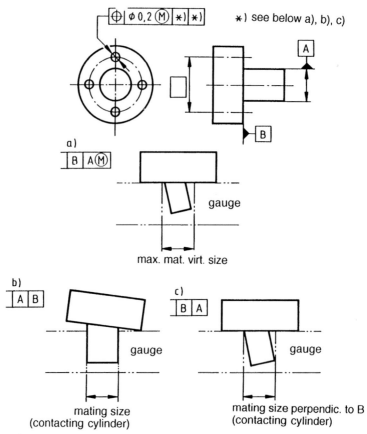

Figure 20.36: Sequence and maximum material requirement for datums.

according to the minimum-rock requirement. The surface B contacts the gauge at one point only. (In the case of the positional tolerance of the holes this has no influence on the position.)

In case (c) the surface B must contact the gauge according to the minimum-rock requirement. In this orientation the gauge must enclose the datum feature A with the least possible diameter.

Figure 20.37 and 20.38 show positionally toleranced holes with and without application of the maximum material requirement for the datums.

According to Figure 20.37, the surface A must contact the gauge according to the minimum-rock requirement. The datum feature B must be enclosed by the

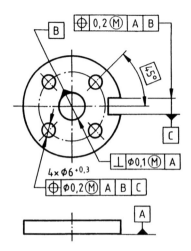

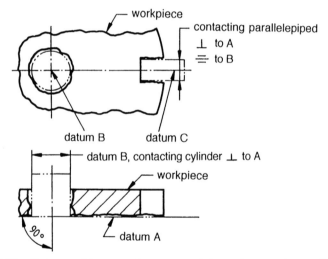

Figure 20.37: Datums without maximum material requirement (regardless of feature size).

maximum possible cylinder perpendicular to the gauge surface A. The slot C must be enclosed by the maximum possible parallelepiped perpendicular to the gauge surface A. The median plane of the parallelepiped contains the axes of the gauge cylinder B.

325

According to Figure 20.38 also, surface A must contact the gauge according to the minimum-rock requirement. The gauge cylinder B and the gauge parallelepiped C are perpendicular to the gauge surface A, and the symmetry plane of the gauge parallelepiped contains the axis of the gauge cylinder B. However, the gauge cylinder B and gauge parallelepiped C have maximum material virtual size.

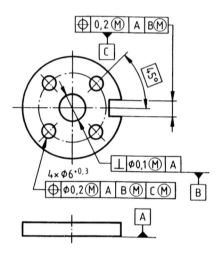

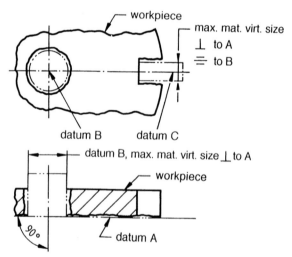

Figure 20.38: Datums with maximum material requirement.

20.7.6 Interchanging toleranced and datum features

Depending on how the datum and toleranced features are chosen, different properties are toleranced. Tolerancing according to Figure 20.39 does not, according to Figure 20.40, limit the perpendicularity deviation of the median face of the tongue in relation to the plane of the hole axes.

See also 18.4

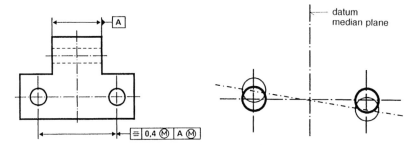

Figure 20.39: Symmetry tolerance and permissible location of the holes; the perpendicularity deviation of the plane of the hole axes in relation to the median face of the tongue is not limited.

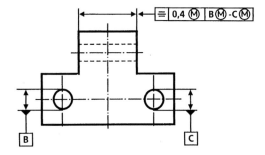

Figure 20.40: Symmetry tolerance also limits the perpendicularity deviation of the symmetry face of the tongue in relation to the plane of the hole axes.

327

20.8   CLEARANCE FIT

20.8.1   Coaxiality of holes

Figure 20.41 shows indications of flatness tolerances according to ISO 1101 on the left independent of each other (the tolerance zones may be different in height level and in orientation) and on the right as common tolerance zone for all surfaces.

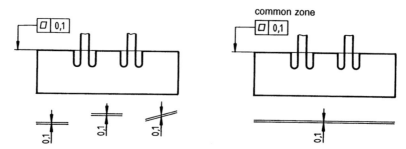

Figure 20.41:   Flatness tolerance zone: (a) independent of each other; (b) common zone.

Figure 20.42 shows similar indications for the straightness of hole axes. It is evident that only the common zone indication limits the coaxiality deviation.

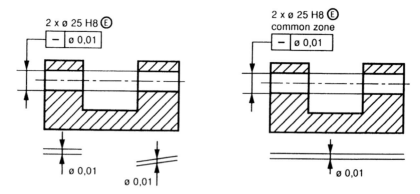

Figure 20.42:   Tolerance zone for the straightness of the axes: (a) independent of each other; (b) common zone. (According to ISO 286: Ø 25 H8 = Ø 25 + 0.033.)

Figure 20.43 shows similar indications for the envelope requirement of holes. Only the indication "common zone" expresses the requirement that both actual holes must not infringe the imaginary envelope cylinder (boundary) of maximum material size that extends over both holes. Without the indication "common zone", each hole separately must respect its envelope cylinder. These envelope cylinders need not be coaxial.

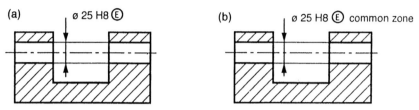

Figure 20.43:   Envelope requirement for holes drawn on the same centre line: (a) independent; (b) common.

Figure 20.44 and 20.45 show the influence on the measured value by chosing the datum. On the right in both figures the measured value on the same actual workpiece is shown. The figures reveal that, on interchanging datum and toleranced features on the same actual workpiece, quite different measured values may arise.

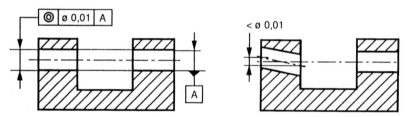

Figure 20.44:   Coaxiality tolerance and measured value.

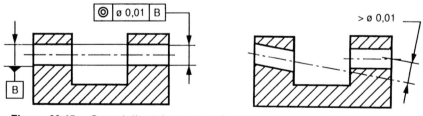

Figure 20.45:   Covaxiality tolerance and measured value with interchanged datum and toleranced feature compared with Figure 20.44.

Examples of Geometrical Tolerancing

Often (e.g. with clearance fits) quite different functional requirements apply (e.g. the parts should fit into each other). The requirement is then that both axes are to be contained in a common tolerance cylinder ($\varnothing$ 0.01 in Figure 20.46). This corresponds to gauging with a plug gauge that extends over both holes simultaneously, the diameter of the plug gauge being the maximum material size minus straightness tolerance (maximum material virtual size = $\varnothing$ 24.99).

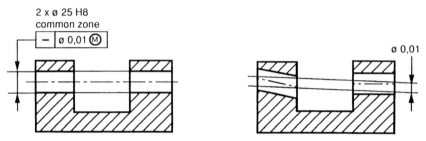

Figure 20.46: Tolerancing of a clearance fit with common pin. (According to ISO 286: $\varnothing$ 25 H8 = $\varnothing$ 25 + 0.033).

Figure 20.47 shows the possibilities for tolerancing of coaxial holes using coaxiality tolerances.

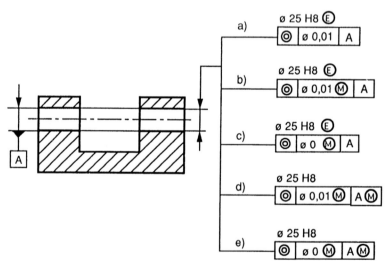

Figure 20.47: Coaxial holes: possibilities of indicating coaxiality tolerances. (According to ISO 286, $\varnothing$ 25 H8 = $\varnothing$ 25 + 0.033.)

In case (a) each hole must respect the envelope requirement Ⓔ separately. The actual axis of the right hole must be contained in a tolerance cylinder of ⌀ 0.01 which is coaxial with the datum axis A. The latter is positioned according to the minimum-rock requirement. The right hole may have a coaxiality deviation of 0.005 (corresponding to a coaxiality tolerance of 0.01; see 18.6).

In case (b) each hole must respect the envelope requirement Ⓔ separately. The actual right hole must respect the maximum material virtual cylinder (gauge) of diameter given by the maximum material size minus the coaxiality tolerance (⌀ 25 − 0.01 = ⌀ 24.99) that is coaxial with the datum axis A. The latter is positioned according to the minimum-rock requirement.

In case (c) each hole must respect the envelope requirement Ⓔ separately. The actual right hole must respect the maximum material virtual cylinder (gauge) of diameter given by the maximum material size minus the coaxiality tolerance (⌀ 25) that is coaxial with the datum axis A. The latter is positioned according to the minimum-rock requirement.

In case (d) a stepped plug gauge must fit in the holes, whose diameter are given by (right) the maximum material size minus the coaxiality tolerance (⌀ 25 − 0.01 = ⌀ 24.99) and (left) the maximum material size (25).

In case (e) a plug gauge must fit simultaneously in both holes, whose diameter is the maximum material size (25).

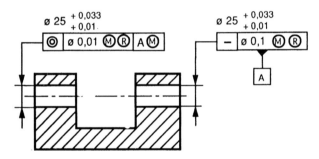

Figure 20.48: Coaxial holes: reciprocity requirement applied.

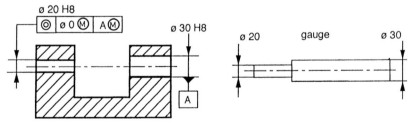

Figure 20.49: Coaxiality of holes with different nominal sizes: tolerancing of clearance fit and relevant gauge.

Figure 20.48 shows a case similar to Figure 20.47 (e); however, the total tolerance (0.033) is distributed on a size tolerance (0.022) and a geometrical tolerance (0.01) as a recommendation. The requirement is for size and geometry to respect the maximum material virtual condition of ∅ 25, i.e. a plug gauge of ∅ 25 must fit simultaneously in both holes.

Figure 20.49 shows the tolerance indication and the relevant gauge when the holes have different nominal sizes.

Figure 20.50 shows the possibilities for tolerancing of coaxial holes using common tolerance zones for the straightness deviation of the axes of the holes or common zones for the envelope requirement.

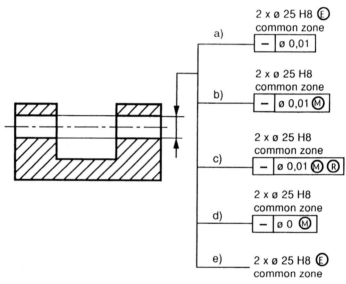

Figure 20.50: Coaxiality of holes: possibilities for indicating common zones. (According to ISO 286: ∅ 25 H8 = ∅ 25 + 0.033.)

In case (a) each hole must respect the envelope requirement Ⓔ separately. The actual axes of the holes must be contained within a common tolerance cylinder of ∅ 0.01. The holes may have a coaxiality deviation of 0.01 (corresponding to a common straightness tolerance zone of ∅ 0.01) regardless of the hole sizes, i.e. also when the holes have maximum material sizes (∅ 25).

In case (b) a plug gauge must fit simultaneously in both holes, whose diameter is given by the maximum material size minus the straightness tolerance (∅ 25 − 0.01 = ∅ 24.99). However, the actual local sizes of the holes must be between ∅ 25 and ∅ 25.033 (= ∅ 25 H8).

In case (c) a plug gauge must fit simultaneously in both holes, whose diameter is given by the maximum material size minus the straightness tolerance ($\emptyset$ $25 - 0.01 = \emptyset$ 24.99). The actual local sizes of the holes must be between $\emptyset$ 24.99 and $\emptyset$ 25.033.

In case (d) a plug gauge must fit simultaneously in both holes, whose diameter is the maximum material size ($\emptyset$ 25).

In case (e) the same applies as in case (d).

Figure 20.51 shows a hinge frame with holes $\emptyset$ 8 D 10 ($+0.098 + 0.040$) that should fit together with a bolt $\emptyset$ 8 h 9 ($0 - 0.036$) without constraint (clearance fit). Figure 20.51 (bottom) shows the extreme case when all actual sizes are at the worst limits. Even then, the parts can still fit without constraint.

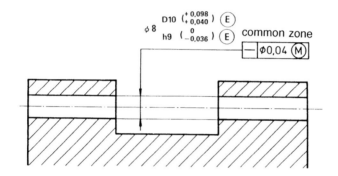

extreme case: max. mat. condition:

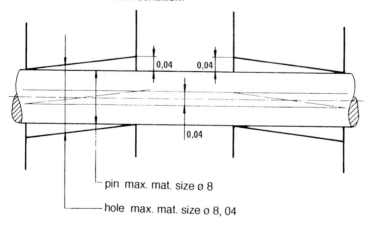

Figure 20.51: Tolerancing of a hinge: clearance fit. (According to ISO 286: $\emptyset$ 8 D10 = $\emptyset$ 8 + 0.098/ + 0.04; $\emptyset$ 8 h8 = $\emptyset$ 8 − 0.036.)

Examples of Geometrical Tolerancing

The holes may be inspected with a plug gauge whose diameter is given by the maximum material size minus the straightness tolerance ($\varnothing$ 8.04 − 0.04 = $\varnothing$ 8). The plug gauge must extend over both holes simultaneously.

Figure 20.52–20.55 show various possibilities for tolerancing of coaxial holes in relation to a supporting surface (e.g. a hinge). On the right the relevant inspection or gauge is shown.

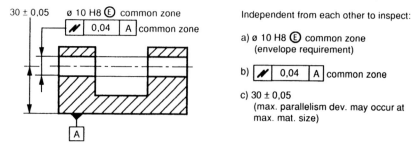

Independent from each other to inspect:

a) ø 10 H8 Ⓔ common zone
   (envelope requirement)

b) | ⟋ | 0,04 | A | common zone

c) 30 ± 0,05
   (max. parallelism dev. may occur at max. mat. size)

Figure 20.52: Tolerances independent of each other.

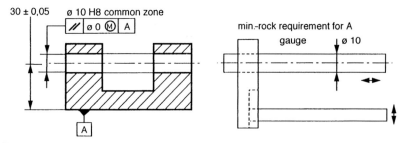

Figure 20.53: Parallelism tolerance contained in the size tolerance of the holes.

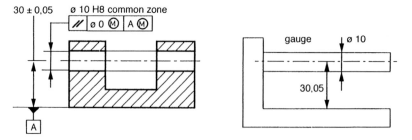

Figure 20.54: Complex fit: tolerance distributed between hole size and distance.

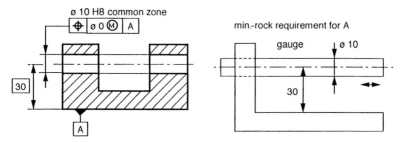

Figure 20.55: Complex fit: tolerance at the hole size only.

Figure 20.56 shows tolerancing of the gap of the hinge to fit with the tongue. The maximum material virtual condition is perpendicular to the datums A and B (minimum-rock requirement).

Figure 20.57 shows tolerancing of four coaxial holes to fit with a bolt. In addition, the holes are positionally toleranced in relation to the datum surfaces A and B.

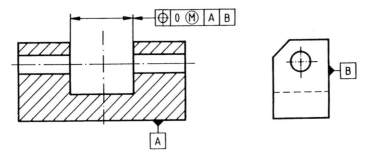

Figure 20.56: Tolerancing of a hinge gap.

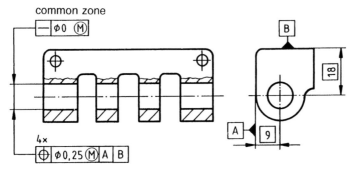

Figure 20.57: Tolerancing of coaxial holes to fit with a bolt.

Examples of Geometrical Tolerancing

20.8.2 Coaxiality of bearing surfaces

Figure 20.58 shows tolerancing of a shaft with a clearance fit and the relevant gauge. If appropriate, the reciprocity requirement, Ⓡ after Ⓡ may be applied additionally (see 9.3.6).

Figure 20.59 shows tolerancing of two bearing surfaces with run-out tolerances related to their common axis. However, the indication of the run-out tolerance excludes the application of the maximum material requirement. Therefore, to ensure the fit, the maximum material sizes of the three cylindrical features are to be decreased by the value of the run-out tolerance. The part should fit into the gauge shown in Figure 20.58. However, problems may arise with lobed forms

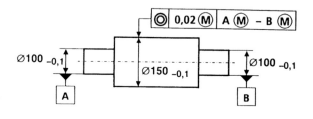

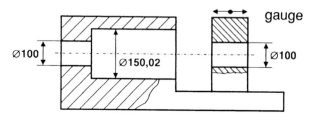

Figure 20.58: Tolerancing of a shaft with clearance fits and the relevant gauge.

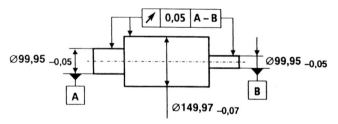

Figure 20.59: Run out tolerances of two bearing surfaces related to their common axis.

when the part is supported in V-blocks during inspection (see 18.7.4.3 and 18.7.10.1.1).

20.8.3 Tolerancing of rings, bushes and hubs for fits

Figure 20.60 shows various possibilities for tolerancing of the perpendicularity deviation of a hole in relation to a plane surface.

In case (a) the hole must respect the envelope requirement Ⓔ. The actual axis of the hole must be contained in a tolerance cylinder of $\varnothing$ 0.05 that is perpendicular to the datum A. The latter is positioned according to the minimum-rock requirement.

In case (b) the hole must respect the maximum material virtual cylinder of diameter given by the maximum material size minus the perpendicularity tolerance ($\varnothing$ 20 − 0.05 = $\varnothing$ 19.95) that is perpendicular to the datum A. The latter is positioned according to the minimum-rock requirement. However, the actual local sizes of the hole must be between $\varnothing$ 20 and $\varnothing$ 20.033.

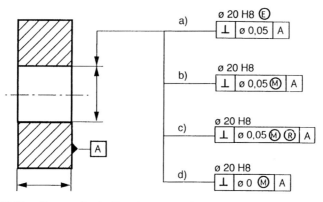

Figure 20.60: Perpendicularity tolerances of a hole related to a plane surface. (According to ISO 286: $\varnothing$ 20 H8 = $\varnothing$ 20 + 0.033.)

In case (c) the hole and its actual sizes must respect the maximum material virtual cylinder of diameter given by the maximum material size minus the perpendicularity tolerance ($\varnothing$ 20 − 0.05 = $\varnothing$ 19.95) that is perpendicular to the datum A. The latter is positioned according to the minimum-rock requirement. Hence the actual local size of the hole may vary from $\varnothing$ 20.033 down to $\varnothing$ 19.95.

In case (d) the hole must respect the maximum material virtual cylinder of diameter given by the maximum material size minus the perpendicularity tolerance ($\varnothing$ 20 − 0 = $\varnothing$ 20) that is perpendicular to the datum A. The latter is positioned according to the minimum-rock requirement.

Figure 20.61 shows various possibilities for tolerancing of the perpendicularity deviations of a hole in relation to two parallel plane surfaces (hole axis in relation to median plane).

In case (a) the hole must respect the envelope requirement Ⓔ. The actual axis of the hole must be contained in a tolerance cylinder of ⌀ 0.05 that is perpendicular to the datum A. The latter comprises two parallel planes a minimum distance apart enclosing the actual surfaces and orientated according to the minimum-rock requirement.

In case (b), the actual hole must respect the maximum material virtual cylinder of diameter given by the maximum material size minus the perpendicularity tolerance (⌀ 20 − 0.05 = ⌀ 19.95) that is perpendicular to the datum A. The latter comprises two parallel planes a minimum distance apart enclosing the actual surfaces and orientated according to the minimum-rock requirement.

In case (c) the actual hole must respect the maximum material virtual cylinder of diameter given by the maximum material size minus the perpendicularity tolerance (⌀ 20 − 0 = ⌀ 20) that is perpendicular to the datum A. The latter comprises two parallel planes a minimum distance apart enclosing the actual surfaces and orientated according to the minimum-rock requirement.

In case (d) the hole must respect the maximum material virtual cylinder of diameter given by maximum material size minus the perpendicularity tolerance (⌀ 20 − 0.05 = ⌀ 19.95) that is perpendicular to the datum A. The latter is established by two parallel planes a maximum material size (25) apart. Between these planes, the actual workpiece may float (translate and rotate). In other words, the workpiece must fit into a gauge of maximum material size minus the

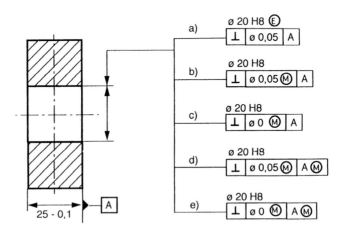

Figure 20.61: Perpendicularity tolerance of a hole in relation to two parallel plane surfaces. (According to ISO 286: ⌀ 20 H8 = ⌀ 20 + 0.0333.)

perpendicularity tolerance at the hole and maximum material size at the width. However, the actual sizes of the hole must be between ∅ 20 and ∅ 20.033.

In case (e) the hole must respect the maximum material virtual cylinder of diameter given by the maximum material size minus the perpendicularity tolerance (∅ 20–0 = ∅ 20) that is perpendicular to the datum A. The latter is established by two parallel planes a maximum material size (25) apart. Between these planes, the actual workpiece may float (translate and rotate). In other words, the workpiece must fit into a gauge of maximum material size minus perpendicularity tolerance at the hole and maximum material size at the width (see Figure 13.4).

Figure 20.62 shows a case similar to Figure 20.61 (d). However, the actual local sizes of the hole may vary between ∅ 20.033 and ∅ 19.95. The reciprocity requirement is applied to give a recommendation as to how to distribute the total tolerance on size and geometry.

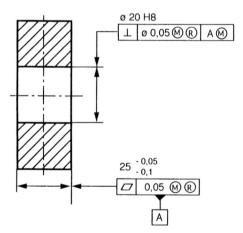

Figure 20.62: Perpendicularity tolerance of a hole in relation to two parallel plane surfaces: reciprocity requirement applied. (According to ISO 286: ∅ 20 H8 = ∅ 20 +0.033.)

339

20.8.4 Rectangle fit

Figure 20.63 shows tolerancing of rectangular features that should fit together.

Without altering the functional requirement shown in Figure 20.63, it is possible to indicate a manufacturing recommendation as to how the total tolerance on size and geometry can be distributed (reciprocity requirement) (Figure 20.64; see also 9.3.6).

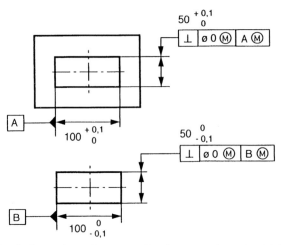

Figure 20.63: Tolerancing of rectangular features that should fit together: total tolerance at the size.

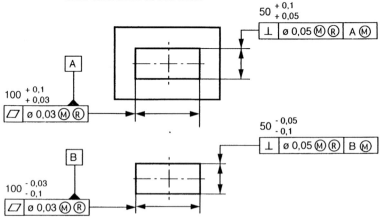

Figure 20.64: Tolerancing of rectangular features that should fit together: geometrical tolerance and size tolerance separated as a recommendation.

20.8.5 Hexagon fit

Figure 20.65 shows a hexagon fit with a minimum clearance 0.1. The total tolerance is indicated at the nominal size.

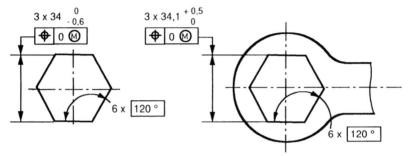

Figure 20.65: Hexagon fit with minimum clearance 0.1: total tolerance at the size.

Figure 20.66 shows the same fit. However, the total tolerance is distributed on size and geometry as a manufacturing recommendation (reciprocity requirement) (see also 9.3.6). Sizes and geometry must respect the maximum material virtual condition: 34 for the head and 34.1 for the wrench, which are in the theoretical exact locations (120°) relative to each other.

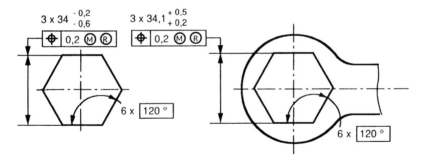

Figure 20.66: Hexagon fit with minimum clearance 0.1: total tolerance distributed on size and geometry as a recommendation (reciprocity requirement).

Figure 20.67 shows a similar fit using positional tolerances for the single plane surfaces.

Examples of Geometrical Tolerancing

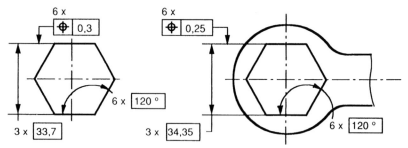

Figure 20.67: Hexagon fit with minimum clearance 0.1: positional tolerances for the single plane surfaces.

Figure 20.68 shows the effect of these tolerances. The limitation at the maximum material limit (go gauge) is the same as according to Figures 20.65 and 20.66. At the minimum material limit the geometrical ideal form and orientation apply according to Figure 20.67, but not according to Figures 20.65 and 20.67. The flatness deviation may occur up to 0.6 or 0.5 respectively (= size tolerance)

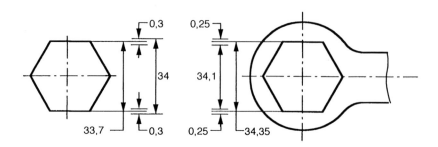

minimum clearance 34,1 - 34 = 0,1

Figure 20.68: Effect of positional tolerancing according to Figure 20.67.

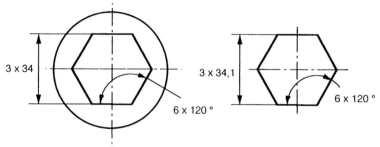

Figure 20.69: Go gauges for D&T according to Figures 20.65–20.67.

342

according to Figures 20.65 and 20.66, and 0.3 or 0.25 respectively according to Figure 20.67.

Figure 20.69 shows the go gauges for D&T according to Figures 20.65–20.67.

20.8.6 Cross-fit: side faces only

Figure 20.70 shows a cross-fit composed of eight side faces of each part (the four head faces have larger clearances, and are not included in the fit). The reciprocity requirement may also be applied (see Figure 20.66).

Figure 20.71 shows a similar fit using positional tolerances for the single plane surfaces. The limitation at the maximum material limit (go gauge) is the same as according to Figure 20.70. At the minimum material limit the geometrical ideal form and orientation apply according to Figure 20.71 but not according to Figure 20.70. Figure 20.72 shows the go gauges for D&T according to Figures 20.70 and 20.71.

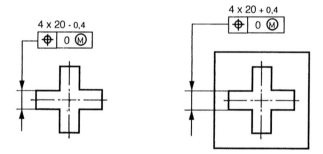

Figure 20.70: Cross-fit: total tolerance at the size.

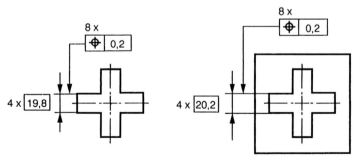

Figure 20.71: Cross-fit: positional tolerances for the single side faces.

Examples of Geometrical Tolerancing

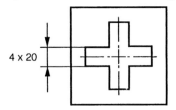

Figure 20.72: Go gauges for D&T according to Figures 20.70 and 20.71.

20.8.7 Cross-fit: all around

Figure 20.73 shows a cross-fit covering all faces around. The reciprocity requirement may also be applied (see Figure 20.66).

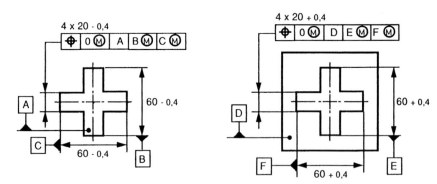

Figure 20.73: Cross-fit: total tolerance at the size.

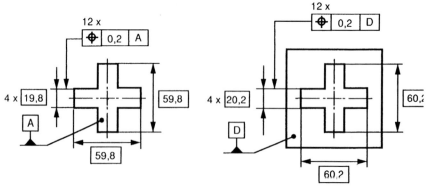

Figure 20.74: Cross-fit: positional tolerances for the single plane surfaces.

344

Figure 20.74 shows a similar fit using positional tolerances for the single plane surfaces.

Figure 20.75 shows the go gauges for D&T according to Figures 20.73 and 20.74.

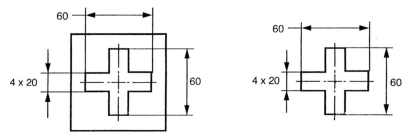

Figure 20.75: Go gauges for D&T according to Figures 20.73 and 20.74.

20.8.8 Splines

Figure 20.76 shows a spline fit. The total tolerance is indicated at the nominal size. Figure 20.77 shows the same fit, but the total tolerances are distributed on size and geometry as a manufacturing recommendation (reciprocity requirement) (see 9.3.6). Figure 20.78 shows a similar fit using positional tolerances for the single plane surfaces. In all three cases the same go gauge applies (gauge sizes 5 and $\varnothing$ 30) (Figure 20.79).

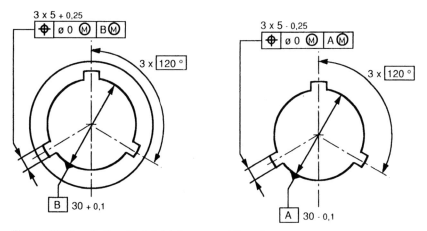

Figure 20.76: Spline fit: total tolerance at the size.

Examples of Geometrical Tolerancing

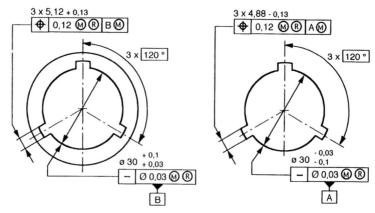

Figure 20.77: Spline fit: total tolerances distributed on size and geometry (reciprocity requirement).

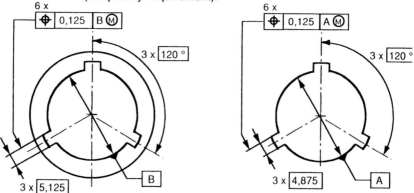

Figure 20.78: Spline fit: positional tolerances for the single plane surfaces.

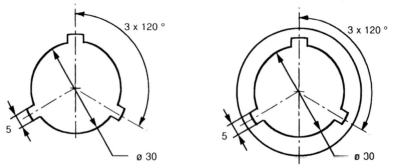

Figure 20.79: Go gauges for D&T according to Figures 20.76–20.78.

20.8.9 Plug-and-socket fit

Figure 20.80 shows a plug-and-socket fit. The total tolerance is indicated at the nominal size of the cylindrical feature.

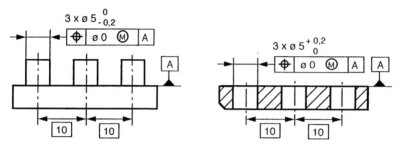

Figure 20.80: Plug-and-socket fit: total tolerance at the size.

Figure 20.81 shows the same fit, but the total tolerance 0.2 is distributed on size and position as a manufacturing recommendation (reciprocity requirement) (see 9.3.6).

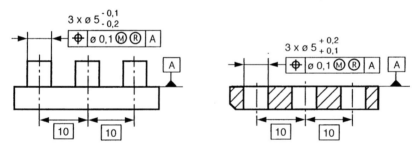

Figure 20.81: Plug-and-socket fit: total tolerance distributed on size and position (reciprocity requirement).

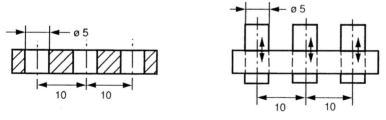

Figure 20.82: Go gauge for D&T according to Figures 20.80 and 20.81.

Figure 20.82 shows the go gauges for D&T according to Figures 20.80 and 20.81. Because the positional tolerance is related to the primary datum A, the workpiece surface A must be orientated according to the minimum-rock requirement relative to the gauge surface A.

Figure 20.83 shows a plug-and-socket fit where the socket has a frame and forms a further fit with the plug circumference, which is related to the fits of the pins and holes.

Figure 20.84 shows the go gauges for D&T according to Figure 20.83.

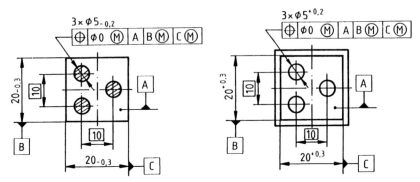

Figure 20.83: Plug-and-socket fit with frame fit: total tolerance at the sizes.

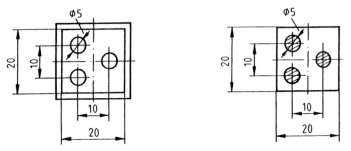

Figure 20.84: Go gauges for D&T according to Figure 20.83.

Figure 20.85 shows tolerancing of a socket similar to that shown in Figure 20.80, but with additional positional tolerances related to datums A, B and C. The holes must respect the maximum material virtual cylinder of $\emptyset$ 20 (perpendicular to datum A and 40 apart) and the maximum material virtual cylinder of $\emptyset$ 19.5, which are in the theoretical exact orientation and location in relation to the datums A, B and C.

Figure 20.86 shows the go gauges for D&T according to Figure 20.85. The workpiece surface A must be orientated according to the minimum-rock requirement relative to the surface A of gauge 1a; then the gauge 1b must fit in

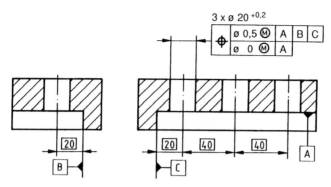

Figure 20.85: Tolerancing of a socket (similar to Figure 20.80) relative to a datum system: tolerances for the position of the holes in relation to each other within the size tolerance.

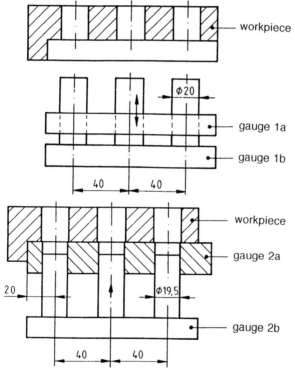

Figure 20.86: Separated go gauges (go gauge set) for D&T according to Figure 20.85.

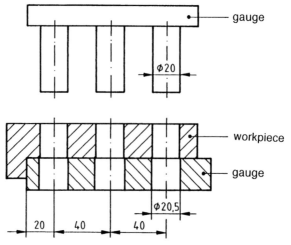

Figure 20.87: Combined go gauge for D&T according to Figure 20.85.

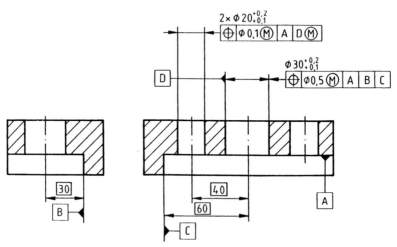

Figure 20.88: Tolerancing of a socket, relative to a datum (guiding) hole A that is toleranced relative to a datum system.

the workpiece. Then the workpiece must be positioned according to the minimum-rock requirement relative to the surfaces A and B and relative to the surface C of gauge 2a (see 3.4); then the gauge 2b must fit in the workpiece.

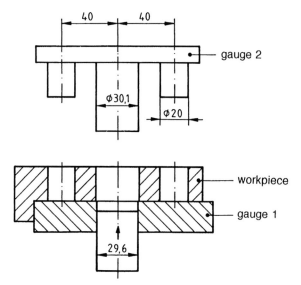

Figure 20.89:   Separated go gauges (go gauge set) for D&T according to Figure 20.88.

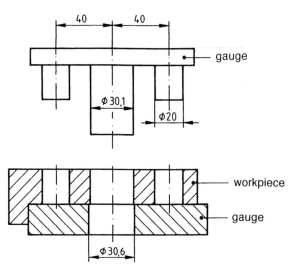

Figure 20.90:   Combined go gauge for D&T according to Figure 20.88.

Examples of Geometrical Tolerancing

Both gauging procedures may be combined. Figure 20.87 shows the appropriate gauge. The holes in the gauge are greater than the maximum material size by the amount of the positional tolerance, which is related to A, B and C. However, this gauging does not inspect the perpendicularity requirement of 0 Ⓜ related to A.

Figure 20.88 shows positional tolerancing of two holes related to a datum hole D. Datum hole D is related to a datum system A, B, C.

Figure 20.89 shows the appropriate go gauge set.

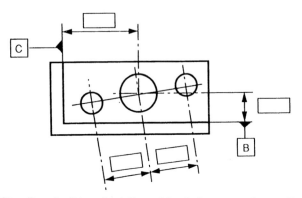

Figure 20.91: Permissible orientation of the holes according to Figure 20.88.

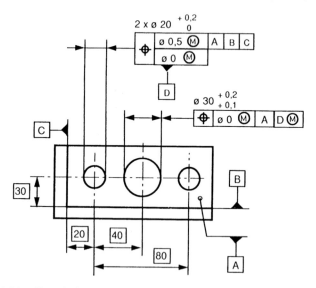

Figure 20.92: Two holes positionally toleranced in relation to the datum system A,B,C in order to limit the effect of Figure 20.91.

352

Both gauging procedures may be combined. Figure 20.90 shows the appropriate gauge. The gauge plug diameter for the datum hole D is equal to the maximum material size. The hole in gauge part 1 has maximum material size plus positional tolerance. The workpiece must be positioned according to the minimum-rock requirement relative to the surfaces A, B and C of gauge part 1 (see 3.4). However, this gauging does not inspect the perpendicularity requirement 0 Ⓜ related to A.

Tolerancing according to Figure 20.88 does not limit the inclination of the hole pattern in relation to the datums B and C (Figure 20.91).

In order to limit the effect shown in Figure 20.91, at least two holes must be positionally toleranced in relation to the datum system A, B, C (Figure 20.92).

20.8.10 Synopsis of tolerancing of clearance fits

Figure 20.93 shows a synopsis of frequent cases of clearance fits toleranced appropriately for the function.

Figure 20.94 shows a synopsis of frequent functional cases of coaxial holes with clearance fits. The relevant gauges demonstrate the function. In the upper example the indications (a) and (b) have the same meaning.

Examples of Geometrical Tolerancing

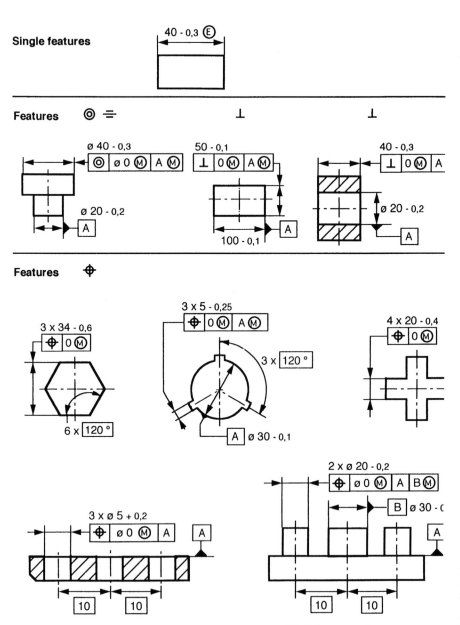

Figure 20.93: Clearance fits: functional cases $\textcircled{E}$, $\textcircled{O}$, $\equiv$, $\perp$, $\oplus$.

354

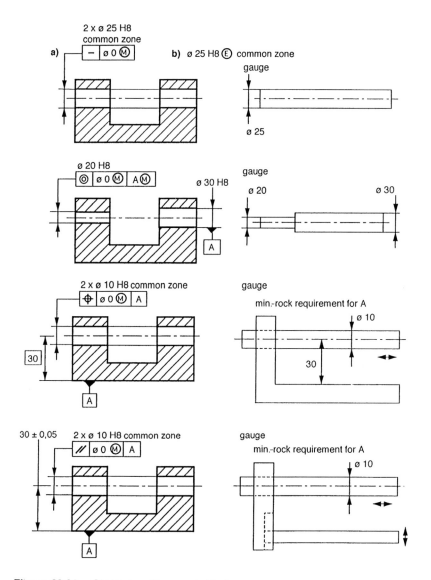

Figure 20.94: Clearance fits: coaxial holes.

20.9 INTERFERENCE FITS AND KINEMATICS

20.9.1 Coaxial holes

Figure 20.95 shows a hinged frame with holes $\varnothing$ 8 N 9 (0 – 0.036) Ⓔ that should fit together with a bolt $\varnothing$ 8 h 9 (0 – 0.036) Ⓔ (transition fit).

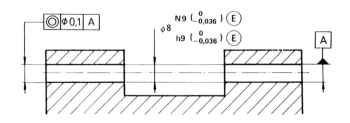

extreme case:

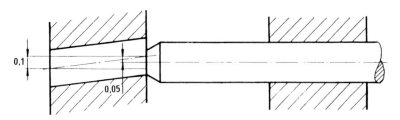

Figure 20.95: Tolerancing of a hinge: transition fit.

Figure 20.95 (bottom) shows the extreme case when all actual sizes are at the worst limits. In this case the coaxiality deviation of the two holes is 0.05. The bolt is guided without clearance by the right hole, and must deflect when entering the left hole. Here the indication of the maximum material requirement Ⓜ would be detrimental, because it would result in a greater deflection of the bolt necessary for the fit. The maximum material requirement would allow greater coaxiality deviations with larger holes, although the bolt may already be guided in one hole without clearance.

Inspection can be performed with two mandrels that fit into the holes without clearance.

Compare with Figure 20.51.

20.9.2   Geometrical ideal form at least material size

Figure 20.96 shows tolerancing when the geometrical ideal form (here a cylinder) at least material size (here the maximum size) must be respected. The surface must not encroach upon this boundary in the direction to the material.

Figure 20.97 shows tolerancing when, in addition to the geometrical ideal boundary at least material size, the envelope requirement Ⓔ should apply (i.e. when the surface must be contained between the geometrical ideal boundaries at maximum material size and at least material size, e.g. for interference fits).

Figure 20.98 has a similar meaning as Figure 20.97. In Figure 20.98 the boundaries are coaxial, while in Figure 20.97 not necessarily so.

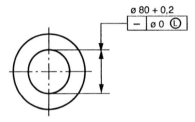

Figure 20.96:   Hole with minimum material requirement.

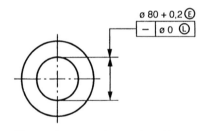

Figure 20.97:   Hole with envelope requirement and least material requirement.

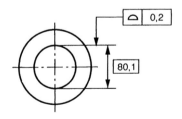

Figure 20.98:   Hole with surface profile tolerance; geometrical ideal forms with maximum and least material sizes must be respected.

20.9.3 Gears

Figure 20.99 shows tolerancing of a gear. Because of a clearance fit, the hole and endfaces are determined through the maximum material requirement. Because of kinematic considerations at the tooth mesh contact, the maximum material requirement applied to the pitch diameter would be detrimental, resulting in greater offset of the tooth contact points.

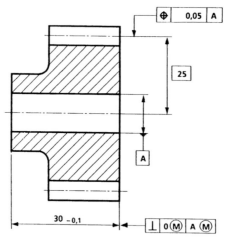

Figure 20.99: Tolerancing of a gear.

20.10 DISTANCES AND THICKNESSES

20.10.1 Spacings

According to Figure 20.100, the median faces of the slots must not deviate by more than ± 0.025 from the symmetrical position in relation to the axis of the datum hole B and the theoretical exact spacing. The positional tolerance is independent of the actual sizes of the slots and of the hole.

According to Figure 20.101, slots and hole must fit into a geometrical ideal counterpart (go gauge) of maximum material size at the hole and maximum material size (3.4) at the slots. With this requirement the positional tolerance depends on the actual sizes and actual forms of the slots and of the hole. The actual size of the slot may be 3.4 when the slot is in the geometrical ideal form and position.

According to Figure 20.102, the position of the right faces of the slots in relation to the axis of the datum hole B is closely toleranced (0.02). The widths of the slots are less so (± 0.05). The positional tolerances of the right slot faces are independent

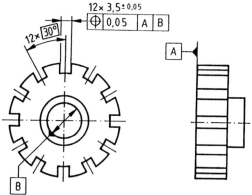

Figure 20.100: Spacing: positional toleranced slot median face.

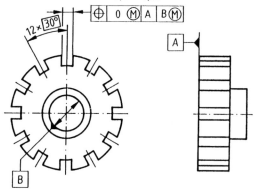

Figure 20.101: Spacing: positional toleranced slot median face with maximum material requirement.

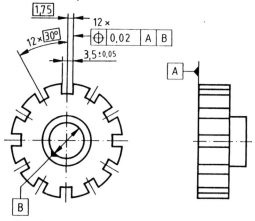

Figure 20.102: Spacing: positionally toleranced slot face.

of the actual sizes of the slots and of the hole.

According to Figure 20.103, the slot faces must respect the geometrical ideal least material virtual condition (boundary 4.05) whose median plane contains the axis of the datum hole B. 'Respect' here means that the boundary lies entirely within the material. The positional tolerance is 0.5 at least material size (3.55), and becomes larger with smaller slots. The positional tolerance is independent of the actual size of the hole.

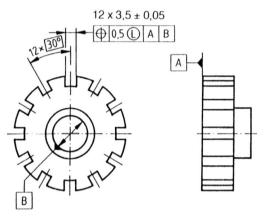

Figure 20.103: Spacing: positionally toleranced slot median face with least material requirement.

20.10.2 Hole edge distance

Figure 20.104 shows (left) tolerancing of distances by linear tolerances and (right) the explanation. The shortest distances from the hole axes to the side faces are toleranced.

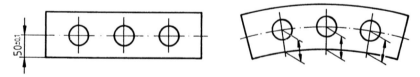

Figure 20.104: Tolerancing of hole face distances by linear tolerances.

Figure 20.105 shows (left) tolerancing of distances by positional tolerances related to a datum face A and (right) the explanation. The distances from the hole axes to the datum, i.e. practically to a contacting gauge plate (simulated datum feature), are toleranced.

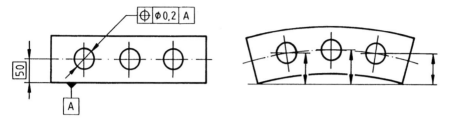

Figure 20.105:   Tolerancing of hole face distances by positional tolerances related to a datum plane.

20.10.3   Equal thickness

Some functions (e.g. of discs of multiple-disc clutches) require small limits on the variation of the actual local sizes (e.g. of thickness) within the same feature (e.g. two parallel plane faces) of a single workpiece. The size deviations of different workpieces, however, may vary over a larger range. The parallelism tolerance is not approriate for this purpose, since a workpiece may have a considerable parallelism deviation but almost no variation in its actual local sizes (Figure 20.106).

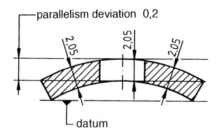

Figure 20.106:   Workpiece with considerable parallelism deviation but almost no variation in its actual local sizes.

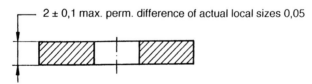

Figure 20.107:   Tolerancing of equal thickness.

Examples of Geometrical Tolerancing

A symbol for equal thickness is not yet standardized internationally. Therefore the following indication may be used: "max. permissible difference of actual local sizes..." (Figure 20.107).

At present it is under discussion whether to standardize the drawing indication for equal thickness as, for example, 2 ± 0.1 Δ 0.05.

20.10.4 Minimum wall thickness

When the function requires a minimum wall thickness, and the wall thickness and position of the wall depend on each other, then the permissible positional deviation of the wall becomes larger the more the wall deviates from its least material size (the more it becomes thicker).

Figure 20.108 shows tolerancing to ensure a minimum wall thickness (2.705) between hole and recess. When the hole is at its least material size (∅ 4.2), its axis must be contained in a tolerance zone ∅ 0.25. In the worst case, when the hole axis has the largest positional deviation, the remaining wall thickness is 2.705 (Figure 20.109 left). When the hole is at its maximum material size (∅ 3.95), the tolerance zone of the hole axis increases by the amount of the size tolerance (∅ 0.25). Then also, in the worst case the remaining wall thickness is 2.705 (Figure 20.109 right).

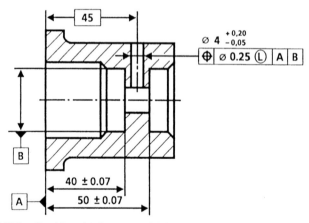

Figure 20.108: Positional tolerance with least material requirement to ensure minimum wall thickness.

In other words, the hole must respect the least material virtual cylinder of diameter given by the least material size plus the positional tolerance (4.2 + 0.25 = 4.45) that is in the theoretical exact position. Between the least material virtual cylinder and the recesses there remains a minimum wall thickness of 2.705 (Figure 20.109).

362

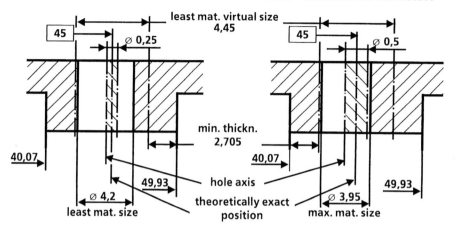

Figure 20.109: Calculation of the minimum wall thickness 2.705 for tolerancing according to Figure 20.108: left, hole with least material size; right, with maximum material size.

Figure 20.110 shows tolerancing to ensure a minimum wall thickness of 2.5. The permissible positional deviation of the inner cylinder increases when the diameter of the inner cylinder decreases. The permissible positional deviation of the outer cylinder increases when the diameter of the outer cylinder increases.

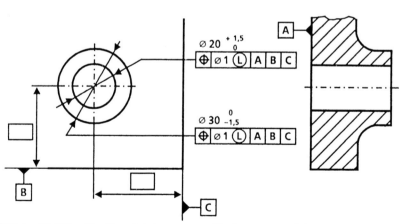

Figure 20.110: Positional tolerances with least material requirements to ensure minimum wall thickness.

Figure 20.111 shows the theoretical exact positions of the least material virtual cylinders that must be respected by the workpiece surfaces. The left side illustrates the extreme case when workpiece features are at least material size, and the right side that when they are at maximum material size.

If appropriate, the reciprocity requirement (ⓁL after the size tolerance) may be applied additionally (see 11.3.4).

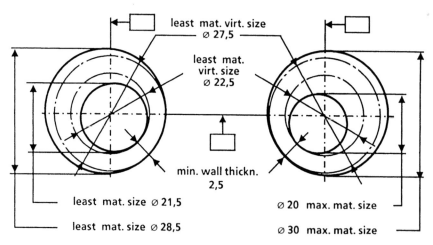

least mat. virt. size
⌀ 27,5

least mat. virt. size
⌀ 22,5

min. wall thickn. 2,5

least mat. size ⌀ 21,5

⌀ 20 max. mat. size

least mat. size ⌀ 28,5

⌀ 30 max. mat. size

Figure 20.111: Calculation of the minimum wall thickness 2.5 for tolerancing according to Figure 20.110: left: features at least material size; right, at maximum material size.

20.11 GEOMETRICAL IDEAL FORM AT MAXIMUM AND LEAST MATERIAL SIZES

In Figure 20.112 (a) the toleranced cylindrical feature should not violate the maximum material virtual condition (limiting boundary) of ⌀ 20. In order to meet this requirement, the manufacturer must distribute the total tolerance of 0.3 on size and form (e.g. 0.2 for size deviations and 0.1 for form deviations). In Figure 20.112 (b) this distribution is already indicated. However, the indications Ⓜ and Ⓡ after the straightness tolerance indicate that the straightness and size tolerances are increased to the extent given by the maximum material virtual condition.

Both indications, Figure 20.112 (a) and (b), specify the same functional

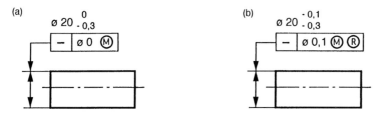

Figure 20.112: Maximum material requirement: without (a) and with (b) reciprocity requirement.

requirement (maximum material virtual condition of ⌀ 20 and actual local sizes within 20 − 0.3). Figure 20.112 (b) gives further information on the recommended distribution of the total tolerance on size and form, i.e. it provides a symbolization for communication between production planning and work shop.

Figure 20.113 (a) and (b) gives analogue information on the least material side. The cylindrical feature should not violate the least material virtual condition (limiting boundary) of ⌀ 19.7. In order to meet this requirement, the manufacturer must distribute the total tolerance of 0.3 on size and form (e.g. 0.2 for size deviations and 0.1 for form deviations). In Figure 20.113 (b) this distribution is already indicated. However, the indications Ⓛ and Ⓡ after the straightness tolerance indicate that the straightness and size tolerances are increased to the extent given by the least material virtual condition.

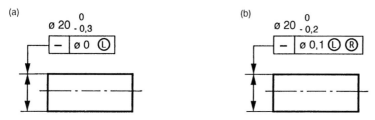

Figure 20.113: Least material requirement: without (a) and with (b) reciprocity requirement.

Figure 20.114 (a) and (b) shows the drawing indications when both limiting boundaries (maximum material virtual condition and least material virtual condition) are to be specified. In order to meet these requirements, the manufacturer must distribute the total tolerance of 0.3 on size and form (e.g. 0.1 for size deviations and 0.1 on each side (maximum material side and least material side) for form deviations). In Figure 20.114 (b) this distribution is already indicated. However, the indications Ⓡ after the symbols Ⓜ and Ⓛ after the straightness tolerance indicate that the straightness and size tolerances are increased to the

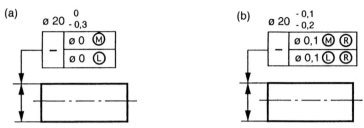

Figure 20.114: Maximum material requirement and least material requirement simultanously; without (a) and with (b) reciprocity requirement.

extent given by the maximum material virtual condition and the least material virtual condition.

The left indications (a) specify the same functional requirements as the right indications (b); for example Figures 20.112 (a) and (b) specify the maximum material virtual condition of $\emptyset$ 20 and actual local sizes to be within $\emptyset$ 20 -0.3. However, the right indications (b) give further information on the recommended (not mandatory) distribution of the total tolerance on size and form.

20.12 FLEXIBLE PARTS

Figure 20.115 shows an example of tolerancing of a flexible part. In the free state, only subject to gravity in the indicated direction, the cylinder B must meet the roundness tolerance of 2.5. Its averaged diameter must be within 1350 ± 0.5 (calculated from at least four equally spaced local sizes or from the circumference). The right cylinder must meet the run-out tolerance of 2 when constrained as specified. Its averaged diameter must be within 1000 ± 0.5 (The average diameter does not depend on whether the free state or constrained condition applies.)

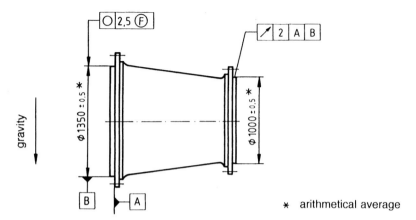

Figure 20.115: Tolerancing of a flexible part. Non-rigid ISO 10 579. Restrained condition: surface A is mounted with 64 bolts M 6x1 with a torque of 9 to 15 N m, and the B feature is restrained to the corresponding maximum material limit.

∗ arithmetical average

21 Differences between ISO Standards and Others

21.1 ANSI Y14.5M, ANSI B89.3.1

21.1.1 Symbols

The National Standard of the United States of America, ANSI Y14.5M, specifies in addition to or deviating from ISO 1101 the symbols and drawing indications shown in Tables 21.1 and 21.2.

21.1.2 Rules

ANSI Y14.5M deviates in some rules from the specifications given in the ISO Standards.
ANSI Y14.5M specifies the following rules.

Rule #1 Where only a tolerance of size is specified, the limits of size of an individual feature prescribe the extent to which variations in its geometrical form, as well as size, are allowed.

In other words, for individual features of size the envelope requirement applies without drawing indication. Exceptions are as follows:

- stock materials (e.g. bars, sheets, tubing and structural shapes) for which straightness, flatness or other geometric tolerances are standardized and which remain in the "as-furnished" condition on the finished part;

- (non-rigid) parts subject to free-state variation in the unrestrained condition (see 12);

- features with a straightness tolerance to the axis or a flatness tolerance to the median face.

Features with straightness tolerances to the generator lines are not exceptions. The envelope requirement applies as well as the straightness tolerance. Both requirements must be respected.
Rule #1 applies only to individual features of size. The rule does not apply to deviations of orientation or location or run-out of one feature related to another.
Rule #1 is similar to the drawing title box indication ISO 2768 ... E and to

369

Table 21.1: Symbols according to ANSI Y14.5M

Symbol	Designation	Interpretation
⌓⁓⁓	ALL AROUND	Profile all around, applicable to the bounded line shown
ALL OVER ⌓	ALL OVER	Everywhere, applicable to all surfaces (e.g. of a casting)*
-A-	Datum	Same as according to ISO 1101 ▶ A
Ⓢ	REGARDLESS OF FEATURE SIZE	According to ANSI Y14.5M, to be indicated with positional tolerances, if applicable According to ISO 1101, not standardized and not to be indicated. The meaning applies if not otherwise specified by Ⓜ or Ⓛ
AVG	AVERAGE	Arithmetical mean (e.g. of the actual sizes of a flexible ring; see Figure 20.115)
Ⓣ	TANGENT	Tolerance applies to the contacting (tangential) element
↔	BETWEEN	Tolerance applies to a limited segment of a surface between designated extremities
CR	CONTROLLED RADIOS	See Figure 21.2
⟨ST⟩	STATISTICAL TOLERANCING	Tolerance is based on statistical tolerancing and shall only be employed where the appropriate statistical process control will be used

*Not specified in ANSI Y14.5M, but usual in US practice.

the German Standard DIN 7167. The difference is that ANSI Y14.5M excludes also those features with a straightness tolerance of the axis indicated that is smaller than the size tolerance (the other standards do not exclude those features).

Table 21.2: Drawing indications according to ANSI Y14.5M

<table>
<tr><th>Drawing indication</th><th>Designation</th><th>Interpretation</th></tr>
<tr><td>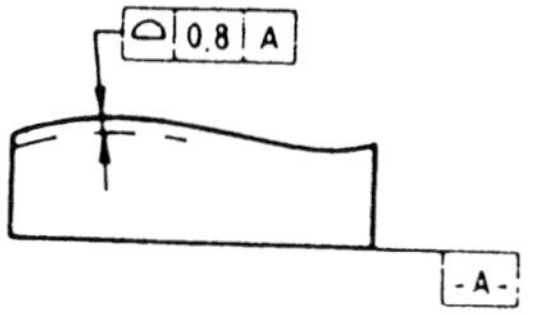
0,8 A
-A-</td><td>UNILATERAL
TOLERANCE
INSIDE</td><td>Geometrical ideal profile
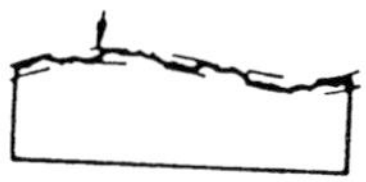</td></tr>
<tr><td>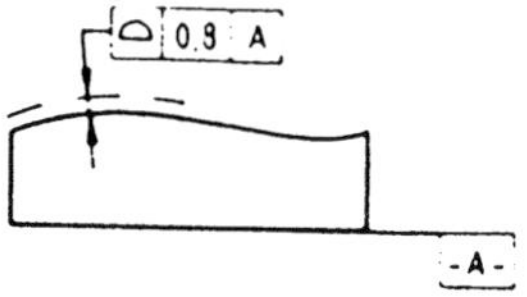
A
-A-</td><td>UNILATERAL
TOLERANCE
OUTSIDE</td><td>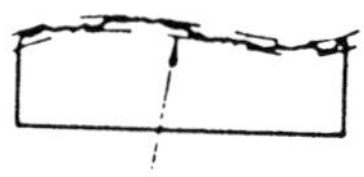
Geometrical ideal profile</td></tr>
<tr><td>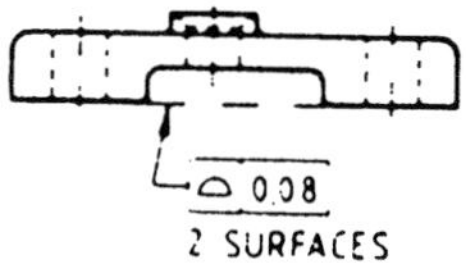
0,08
2 SURFACES</td><td></td><td>ANSI Y14.5M:
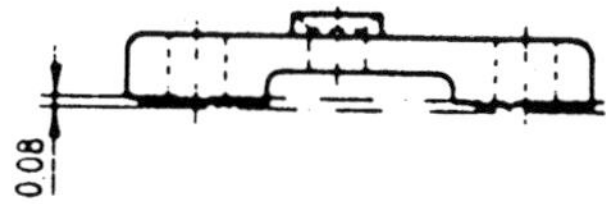
0,08
ISO 1101:
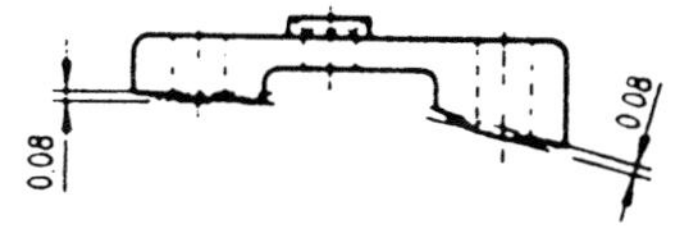
0,08
0,08</td></tr>
<tr><td>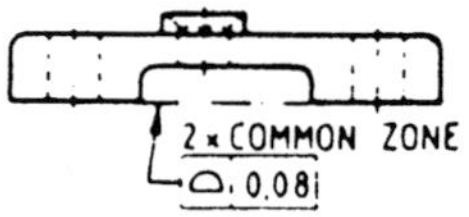
2 x COMMON ZONE
0,08</td><td>COPLANARITY</td><td>ISO 1101 and ANSI Y14.5M:
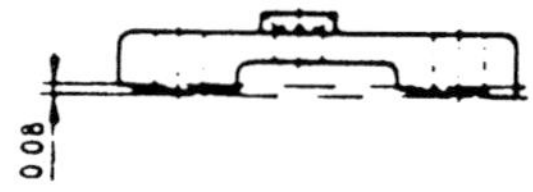
0,08</td></tr>
</table>

Differences between ISO Standards and Others

Rule for radii: A radius symbol R (preceding the dimension) creates a zone defined by two arcs (the minimum and maximum radii). The part surface must lie within this zone, Figure 21.1

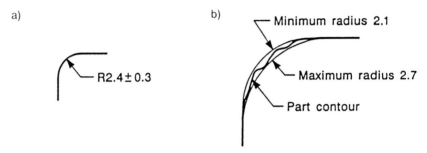

a)

b) Minimum radius 2.1

R2.4 ± 0.3

Maximum radius 2.7

Part contour

Figure 21.1: Radius tolerance, (a) drawing indication, (b) interpretation.

A controlled radius symbol CR (preceding the dimension) creates a tolerance zone defined by two arcs (the minimum and maximum radii) that are tangent to the adjacent surfaces. When specifying a controlled radius, the part contour within the crecent-shaped tolerance zone must be a fair curve without reversals. Additionally, radii taken at all points on the part contour shall neither be smaller than the specified minimum limit nor larger than the maximum limit, see Figure 21.2).

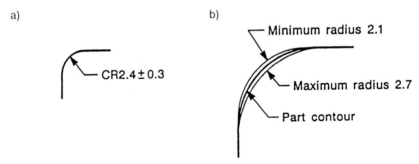

a)

b) Minimum radius 2.1

CR2.4 ± 0.3

Maximum radius 2.7

Part contour

Figure 21.2: Controlled radius tolerance CR, (a) drawing indication, (b) interpretation.

There is not yet a precise definition of the radius tolerance in the ISO-Standards.

Rule for angular surfaces Where an angular surface is defined by a combination of a linear dimension and an angle, the surface must lie within a tolerance zone represented by two non-parallel planes (see Figure 21.2). The tolerance zone will become wider as the distance from the apex of the angle increases. Where a

372

tolerance zone with parallel boundaries is desired, a basic angle may be specified as in Figure 21.3. Additionally, an angularity tolerance may be specified within these boundaries (Figure 21.4).

Only tolerancing according to Figure 21.5 complies with international practice,

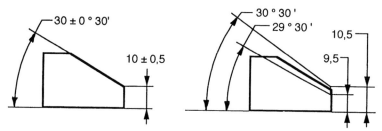

Figure 21.3: Tolerancing of an angular surface using a combination of linear and angular dimensions, according to ANSI Y14.5.

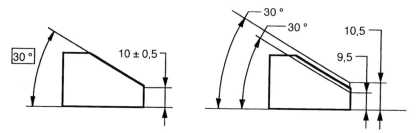

Figure 21.4: Tolerancing of an angular surface with a basic angle.

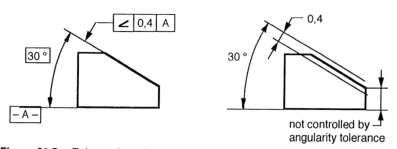

Figure 21.5: Tolerancing of an angular surface with an angularity tolerance.

and is therefore recommended. Tolerancing according to Figure 21.4 is not in accordance with ISO 1101, because datum and angularity tolerance are missing.

373

Differences between ISO Standards and Others

Drawing indications according to Figure 21.3 have a different meaning according to ISO 8015 (see Figure 3.42).

Former rule #2 ANSI Y14.5M - 1982 (positional tolerance rule)   With tolerances of position "regardless of feature size" Ⓢ the maximum material requirement Ⓜ or the least material requirement Ⓛ must be specified on the drawing with respect to the individual tolerance, datum reference, or both, as applicable.

According to a former rule (USASI Y14.5 - 1966 and ANSI Y14.5 - 1973), with positional tolerances, the maximum material requirement Ⓜ was always applied to the toleranced feature as well as to the datum feature, without any indication Ⓜ. This was not in line with the international practice, according to which Ⓢ always applies if not otherwise specified (with Ⓜ or Ⓛ (ISO 8015 and ISO 2692).

New rule #2 ANSI Y 14.5M - 1984   Regardless of feature size applies, with respect to the individual tolerance, datum reference, or both, where no modifying symbol is specified. Ⓜ or Ⓛ must be specified on the drawing where it is required. This new rule # 2 coincides with ISO practices (ISO 1101, ISO 2692, ISO 8015).

New rule #2a ANSI Y 14.5M - 1984 (alternative praxis)   For a tolerance of position Ⓢ **may** be specified on the drawing with respect to the individual tolerance, datum reference, or both, as applicable. This new rule # 2a is not in line with international practice. The symbol Ⓢ does not exist in the ISO-Standards. Regardless of feature size, applies everywhere, if not otherwise specified (with Ⓜ or Ⓛ).

Former rule #3 ANSI Y 14.5M - 1982 (regardless-of-feature-size rule)   For all geometric tolerances, other than positional tolerances, "regardless of feature size" Ⓢ applies, with respect to the individual tolerance, datum reference, or both, where no modifying symbol is specified. This coincides with the International Standards ISO 8015, ISO 2692 and ISO 1101. This rule is now covered by the new rule # 2 ANSI Y 14.5M - 1984.

Pitch diameter rule   Each tolerance of orientation or position (*location*) and datum reference specified for a screw thread applies to the axis of the thread derived from the pitch cylinder. Where an exception to this practice is necessary, the specific feature of the screw thread (such as MINOR DIA or MAJOR DIA) should be stated beneath the feature control frame or the datum feature symbol, as applicable.

Each tolerance of orientation or position (location) and datum reference specified for gears and splines must designate the specific feature to the gear or spline to which each applies (such as MAJOR DIA, PITCH DIA or MINOR

DIA). This information is stated beneath the feature control frame or the datum feature symbol, as applicable.

A similar rule is planned for the next edition of ISO 1101 (following the 1983 edition). The standardized symbols are MD for major, LD for minor (least) and PD for pitch diameter.

Datum virtual condition rule Where a datum feature of size is applied on a MMC basis, machine and gauging elements in the processing equipment, which remain constant in size, may be used to simulate a true geometrical counterpart of the feature and establish the datum. In each case the size of the true geometric counterpart (simulated datum) is determined by the specified MMC limit of size of the datum feature or its MMC virtual condition, where applicable.

Where a primary or single datum feature of size is controlled by a tolerance, the size of the true geometric counterpart used to establish the simulated datum is the MMC limit of size roundness or cylindricity. Where a straightness tolerance is applied on a MMC basis, the size of the true geometric counterpart used to establish the simulated datum is the MHC virtual condition. Where a straightness tolerance is applied on a RFS basis, the size of the true geometric counterpart is the applicable inner or outer boundary, see 21.1.9.

Where secondary or tertiary datum features of size in the same datum reference frame are controlled by a specified tolerance of location or orientation with respect to each other, the size of the true geometric counterpart used to establish the simulated datum is the virtual condition of the datum feature.

Analysis of tolerance controls applied to a datum feature is necessary in determining the size for simulating its true geometrical counterpart. Consideration must be given to the effects of the difference in size between the applicable virtual condition of a datum feature and its MMC limit of size. Where a virtual condition equal to MMC is the design requirement a zero geometrical tolerance at MMC is to be specified (0 Ⓜ).

In other words, for a datum feature of size used as a datum and for a feature of size with a geometrical tolerance applied to its axis or median face, a virtual condition (maximum material virtual condition) applies. The size of the virtual condition depends on the maximum material size and the value of the geometrical tolerance applied to the axis or median face of the datum. When the datum is used as a secondary or tertiary datum, the orientation and location of the virtual condition depend on the primary and secondary datums.

For datum features of size (features with an axis or median plane) with the indication Ⓜ behind the datum letter in the tolerance frame, the following applies.

(a) When for a datum feature with an axis a form tolerance to the feature (not to its axis) is indicated (e.g. roundness tolerance, or straightness tolerance of the generator line), the maximum material virtual size is equal to the maximum material size (Figure 21.6).

When for a datum feature with an axis, a form tolerance to the axis is indicated (e.g. straightness tolerance of the axis), the maximum virtual size is equal to the maximum material size plus or minus the form tolerance of the axis (Figures 21.7–21.8).

(b) When secondary or tertiary datum features of size (with an axis or median plane) occur in the same tolerance frame and are interrelated by a different tolerance of orientation or location between each other, the maximum material virtual condition in the geometrical exact orientation and location relative to each other applies (Figure 21.9). The maximum material virtual size of the datum is equal to the maximum material size plus or minus the corresponding tolerance of orientation or location.

The datum virtual condition rule is not yet internationally standardized. However, part of it is self-evident (see 9.2). What is internationally disputed is the part of the rule dealing with geometrical tolerances to axes without the symbol Ⓜ, (Figure 21.7; see 9.2).

Figure 21.6 shows a case of the rule (a). For the datum B the maximum material requirement Ⓜ applies. In addition, a form tolerance applies to the feature (generator line), not to the feature axis. For the gauge (to inspect the position of the four holes) at the datum B the virtual condition is equal to the maximum material condition (size ∅ 10). This coincides with international practice.

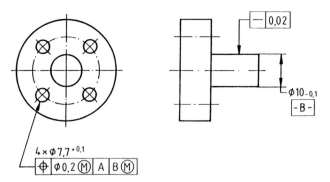

Figure 21.6: Datum feature (B) with an axis and with a form tolerance that does not apply to the latter.

Figure 21.6 shows another case of the rule (a). For the datum B the maximum material requirement Ⓜ applies. In addition, a straightness tolerance of the axis (without Ⓜ) applies. For the gauge (to inspect the position of the four holes) at the datum B the maximum material virtual condition of ⌀ 10.02 applies. This is internationally disputed.

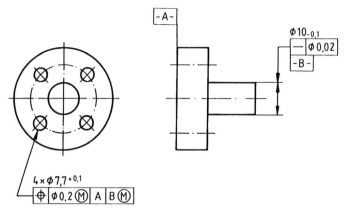

Figure 21.7: Datum feature (B) with an axis and with a straightness tolerance to the latter.

The same maximum material virtual condition of ⌀ 10.02 applies when the straightness tolerance of the axis is ⌀ 0.02 Ⓜ (Figure 21.8). This coincides with international practice.

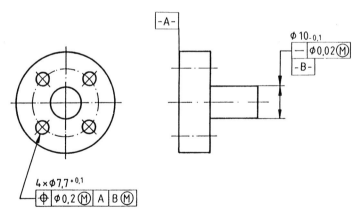

Figure 21.8: Datum feature (B) with an axis and with a straightness tolerance to the latter and with the maximum material requirement to this straightness tolerance.

Figure 21.9 shows a case of the rule (b). The datums A, B and C occur in the same tolerance frame (tolerancing the position of the four holes). The maximum material requirement applies to B and C. B is toleranced in relation to A (with Ⓜ). C is toleranced in relation to A and B (with Ⓜ). For the gauge (to inspect the position of the four holes) at the datums B and C the maximum material

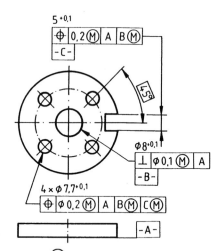

Figure 21.9: Datums with Ⓜ, in the same tolerance frame and toleranced in relation to each other.

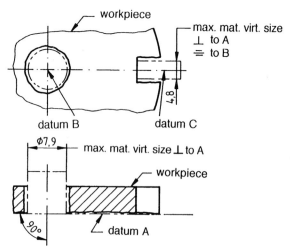

Figure 21.10: Gauge according to Figure 21.7.

378

virtual conditions ($\emptyset$ 7.9 and 4.8) apply (Figure 21.10). This coincides with international practice.

Figure 21.11 shows a part similar to Figure 21.9. However, there is an additional positional tolerance indicated at the datum feature C in relation to the datum D. The latter does not occur in the tolerance frame of the positional tolerance of the four holes. Therefore the tolerance related to D is not to be taken into account for the calculation of the maximum material virtual condition of the datum C of the gauge (to inspect the position of the four holes). The gauge is the same as according to Figures 21.9 and 21.10. This coincides with international practice.

Where a feature of size is applied on a LMC basis, the same applies as according to ISO 2692, see 11.

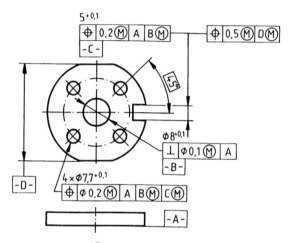

Figure 21.11: Datums with Ⓜ, but in different tolerance frames.

21.1.3 Positional tolerances according to ANSI Y14.5

In the past, according to USASI Y14.5 - 1966 and ANSI Y14.5 - 1973, with positional tolerances, the maximum material requirement was applied to the toleranced features as well as the datum features without any drawing indication of the maximum material requirement. This was applied if not otherwise indicated in the drawing (e.g. by Ⓢ or Ⓛ).

ANSI Y14.5M - 1982 required, with positional tolerancing, the indication Ⓜ, Ⓢ or Ⓛ at the toleranced features (behind the positonal tolerance) as well as at the datum features (behind the datum letter in the tolerance frame).

ANSI Y14.5M - 1994 specifies with the new rule # 2 that it also applies with positional tolerances regardless of feature size, if not otherwise specified with ⓜ or Ⓛ. This coincides with ISO 1101, ISO 2692, ISO 8015.

In the past, according to USASI Y14.5 - 1966 and ANSI Y14.5 - 1973, the following was applied to a positional toleranced pattern (group) of features (e.g. pattern of holes) with adjacend dimensional tolerances (Figure 6.7).

The dimensional tolerances of the adjacent dimensions were applied to the centres of the positional tolerance zones of the pattern (group) of features. Therefore individual positional tolerance zones could be half outside the dimensional tolerance when the feature was at MMC, provided that the positional tolerance zone centres were within the dimensional tolerances. Therefore the actual axes of the holes were allowed to exceed the dimensional tolerances by half of the positional tolerance when the feature was at MMC (Figure 6.7).

According to ISO 8015, this is not permitted. Dimensional tolerances as well as positional tolerances apply to the actual axes of the features.

According to ANSI Y14.5M - 1982, dimensional tolerances for positioning a pattern (group) of features that are positionally toleranced between each other should be avoided and replaced by positional tolerances related to a datum system (Figure 6.8). ANSI Y14.5M - 1994 gives similar advice.

21.1.4 Composite tolerancing and single tolerancing

ANSI Y14.5M - 1994 distinguishes between composite tolerancing and single tolerancing where there are two tolerances of the same type (position or profile) specified for the same features. Composite tolerancing is to be identified by one single tolerance symbol for two tolerances, whereas single tolerancing is to be identified by separate tolerance symbols for each tolerance (Figure 21.12).

In the case of composite tolerancing the upper (always larger) tolerance applies to the **location** of the pattern of features relative to the datums specified, and defines the pattern-locating tolerance zone framework (PLTZF).* The lower (always smaller) tolerance applies to the features relating to each other (interrelation of the features within the pattern) and, if applicable, to the **orientation** of the features as a group relative to the datums specified in the lower tolerance frame, and defines the feature-relating tolerance zone framework (FRTZF).* The theoretical exact dimensions from the datums (used to relate the PLTZF to the datums) do not apply to the smaller tolerance (in the lower tolerance frame), which defines the FRTZF. The actual features (e.g. the actual axes) must lie within both tolerance zones (frameworks) (Figures 21.12a, b).

In the case of single tolerancing each tolerance applies as if no other tolerance were indicated, i.e. the theoretical exact dimensions from the datums specified in

---

*The acronyms are pronounced "Fritz" and "Plahtz".

the lower tolerance frame apply also to the feature-relating tolerance zone framework (FRTZF). The actual features (e.g. the actual axes) must lie within both tolerance zones (framworks) (Figure 21.12c).

ISO 1101 and ISO 5458 do not provide this differentiation. Also, when only one symbol is indicated for two tolerances, the indications of the theoretical exact dimensions relative to the datum(s) apply (as explained above for single tolerancing).

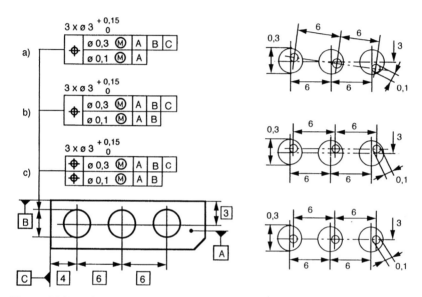

Figure 21.12: Composite positional tolerancing (a, b) and single positional tolerancing (c).

21.1.5 Symmetry tolerances, concentricity tolerances

ANSI Y14.5M - 1982 did not specify the symbol $=$ for symmetry tolerances (it was only contained as former practice). Where it was required that a feature be located symmetrically with respect to the median plane of a datum feature, positional tolerancing (symbol $\oplus$) was to be used. According to ISO 1101, in this case the symbol $=$ is used.

ANSI Y14.5 - 1994 distinguishes between positional tolerances (without modifier Ⓜ or Ⓛ) and concentricity tolerances or symmetry tolerances (without modifier Ⓜ or Ⓛ).

Positional tolerances (symbol ⊕) are applied to the axis or median plane of the actual mating envelope (e.g. maximum inscribed cylinder for holes or minimum circumscribed cylinder for shafts orientated relative to the datum(s)).

Concentricity tolerances (symbol ◎) are applied to the actual axis of the feature (which is composed of the median points of the cross sections of the actual feature).

Symmetry tolerances (symbol ≡) are applied to the actual median points (actual median face) of all opposed or correspondingly located elements of two or more feature surfaces, e.g. the actual median points of two opposite parallel plane surfaces.

ISO 1101 does not provide this distinction. According to ISO 1101 the tolerances of position, concentricity (coaxiality), symmetry, all without modifier, apply (all in the same way) to the actual axis or actual median face of the actual feature, see 3.5.

21.1.6 Tangent plane requirement

ANSI Y14.5M - 1994 introduced the tangent plane requirement to be specified by the symbol ⓣ after the orientation or location tolerance value. Then the tolerance zone applies to the high points of the controlled feature only, not to the entire surface irregularities. For inspection a contacting plate with sufficiently parallel planes may be used (Figure 21.13). At present there is no rule for how to proceed in cases when the plate rocks about the highest points (e.g. when the controlled feature is convex).

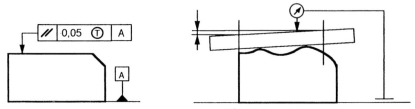

Figure 21.13: Tangent plane requirement ⓣ

21.1.7 Roundness measurements

ANSI B89.3.1 allows the possibility of specifying on drawings for roundness tolerances the measuring conditions, i.e.

- reference method, to be selected from
  - MRS minimum radial separation,
  - LSC least squares circle,
  - MIC maximum inscribed circle,
  - MCC minimum circumscribed circle;

- filter, to be selected from
  0, 1.67, 5, 15, 50, 150, 1500 cycles per revolution;
- stylus tip radius, to be selected from
  0.001, 0.003, 0.010, 0.030, 0.100 in.

See Figure 21.14.

The filters are defined as electrical low-pass filters equivalent to two unloaded *RC* networks in series. Their nominal value (cut-off) corresponds to a sinusoidal frequency (cycles per revolution, cpr) whose amplitude is 70.7% transmitted by the filter.

ISO 4291 specifies a slightly different filter characteristic. The nominal value (cut-off) corresponds to 75% amplitude transmission (see Figure 18.110). However, this difference is practically negligible.

ANSI B89.3.1 states further that, if not otherwise specified (default case), for roundness measurements (roundness tolerances) the following measuring conditions apply:

- reference method   MRS;
- filter   50 cpr;
- tip radius   0.010 in. (0.25 mm).

The ISO Standards do not yet have standardized measuring conditions for the default case. If necessary, they must be agreed upon between parties (e.g. by reference to ANSI B89.3.1).

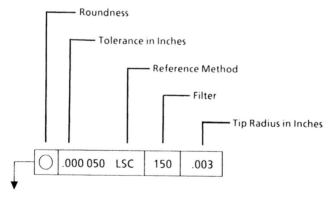

Figure 21.14: Roundness tolerance with measuring conditions according to ANSI B89.3.1.

21.1.8 Multiple patterns of features

According to ANSI Y14.5M positional toleranced multiple patterns of features (two or more patterns of features) located by theoretical exact dimensions relative to the same datum system (common datums in the same order of precedence and modification) are regarded as a single pattern (simultaneous requirement), if not otherwise specified, e.g. with the notation "SEP REQT". This does not apply to the lower segment of composite tolerances unless specific notation is added, e.g. "SIM REQT". ISO-Standards do not specify this expressively, see 20.6.5.

21.1.9 Terminology

Some terms given in ANSI Y14.5 differ from those given by ISO (see Table 21.3).

Table 21.3: Comparison of ANSI Y14.5 and ISO terminologies

ANSI Y14.5M	ISO
Basic dimension	Theoretical exact dimension
Derived median line	Actual axis
Derived median plane	Actual median face
Feature control frame	Tolerance frame
True position TP	Theoretical exact position
Variation	Deviation

ANSI Y14.5M specifies: FIM Full Indicator Movement: the total movement of an indicator when appropriately applied to a surface to measure its variations. The former terms FIR Full Indicator Reading and TIR Total Indicator Reading have the same meaning as FIM.

ANSI Y14.5M defines the terms resultant condition, inner boundary and outer boundary (which are not yet used in ISO Standards) as follows.

Resultant condition: The variable boundary generated by the collective effects of a size feature's specified MMC or LMC material condition, the geometric tolerance for that material condition, the size tolerance, and the additional geometric tolerance derived from the feature's departure from its specified material condition, see Figures 21.15 to 21.18.

Inner boundary: A worst case boundary (that is, locus) generated by the smallest feature (MMC for an internal feature and LMC for an external feature) minus the stated geometric tolerance and any additional geometric tolerance (if applicable) from the feature's departure from its specified material condition, see Figures 21.15 to 21.18.

Outer boundary: A worst case boundary (that is, locus) generated by the largest feature (LMC for an internal feature and MMC for an external feature) plus the stated geometric tolerance and any additional geometric tolerance (if applicable) from the feature's departure from its specified material condition, see Figures 21.15 to 21.18.

These terms may serve to assess the extent of a feature of size relative to its true (theoretical exact) position or to other features.

21.2 BS 308

21.2.1 Tolerancing principles

British Standard BS 308 Part 2:1985 provides both concepts, principle of independency and principle of dependency of size and form with equal status. When the principle of independency is applied, the drawing should be marked with the symbol Ⓐ. Otherwise the principle of dependency applies (see 17.6).

21.2.2 Composite positional tolerancing

When two different positional tolerances are applied to the same features (Figure 6.8), according to BS 308, the larger positional tolerance is regarded as the tolerance of the location of the features as a group (location of the pattern), while the smaller positional tolerance is regarded as the tolerance of the individual locations (location of the features relative to each other). Hence it is specified that the larger positional tolerance applies to the centres of the tolerance zones of the smaller positional tolerance. Therefore the actual axis may exceed the larger tolerance zone by half of the smaller positional tolerance zone (Figure 21.19).

According to ISO 1101 and ISO 5458 (and ANSI Y14.5M), both positional tolerances apply to the actual axes. There is no exceeding of the larger positional tolerance allowed for the actual axis.

When applying ISO 1101, in order to express the same meaning as according to BS 308, the larger positional tolerance is to be modified by the indication "axes of tolerance zones" (Figure 21.20a), or the larger positional tolerance is to be increased by the smaller tolerance value (Figure 21.20b).

21.2.3 Positional tolerances adjacent to dimensional tolerances

In contrast to ISO 1101 and ISO 5458, according to BS 308, the same applies as to ANSI Y14.5 - 1973 (as described in 6.6.1), i.e when there are positional tolerances adjacent to dimensional tolerances, the actual axis may exceed the dimensional tolerance by half the positional tolerance (Figure 6.7).

Differences between ISO Standards and Others

(a)

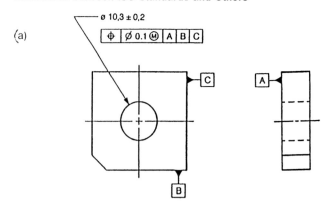

(b)

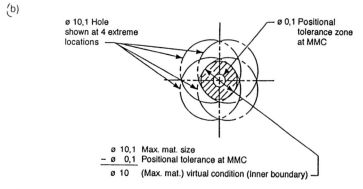

ø 10,1 Max. mat. size
− ø 0,1 Positional tolerance at MMC
ø 10 (Max. mat.) virtual condition (Inner boundary)

c)

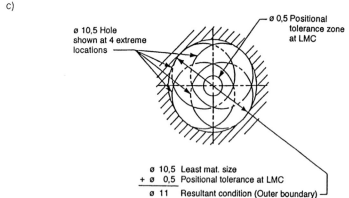

ø 10,5 Least mat. size
+ ø 0,5 Positional tolerance at LMC
ø 11 Resultant condition (Outer boundary)

Figure 21.15: Virtual and resultant condition boundaries, internal feature, MMC (a) drawing indication, (b) virtual condition, (c) resultant condition.

386

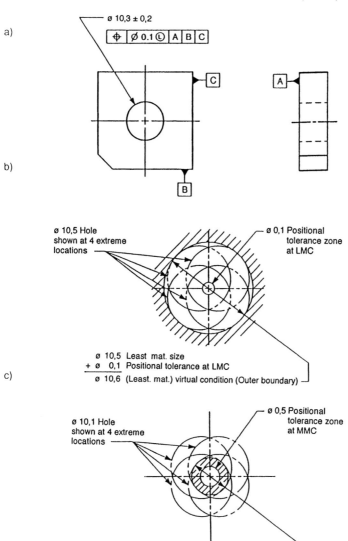

a)

ø 10,3 ± 0,2

⊕ | ø 0.1 Ⓛ | A | B | C

C

A

b)

B

c)

ø 10,5 Hole
shown at 4 extreme
locations

ø 0,1 Positional
tolerance zone
at LMC

    ø 10,5   Least mat. size
+ ø   0,1   Positional tolerance at LMC
    ø 10,6   (Least. mat.) virtual condition (Outer boundary)

ø 10,1 Hole
shown at 4 extreme
locations

ø 0,5 Positional
tolerance zone
at MMC

    ø 10,1   Max. mat. size
− ø   0,5   Positional tolerance at MMC
    ø   9,6   Resultant condition (Inner boundary)

Figure 21.16:   Virtual and resultant condition boundaries, internal feature, LMC (a) drawing indication, (b) virtual condition, (c) resultant condition.

Differences between ISO Standards and Others

a)

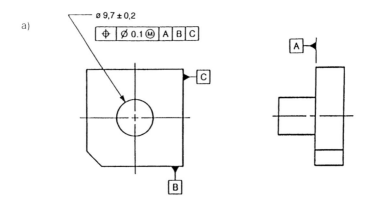

b)

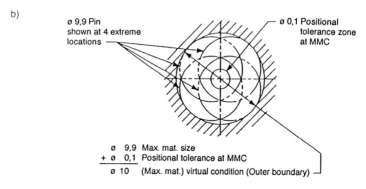

```
ø   9,9  Max. mat. size
+ ø   0,1  Positional tolerance at MMC
  ø  10    (Max. mat.) virtual condition (Outer boundary)
```

c)

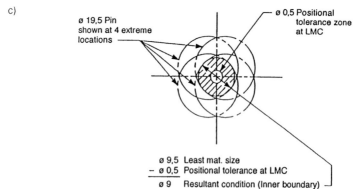

```
  ø 9,5  Least mat. size
− ø 0,5  Positional tolerance at LMC
  ø 9    Resultant condition (Inner boundary)
```

Figure 21.17: Virtual and resultant condition boundaries, external feature, MMC (a) drawing indication, (b) virtual condition, (c) resultant condition.

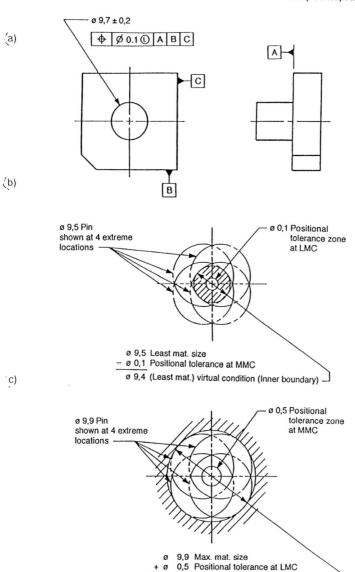

Figure 21.18: Virtual and resultant condition boundaries, external feature, LMC (a) drawing indication, (b) virtual condition, (c) resultant condition.

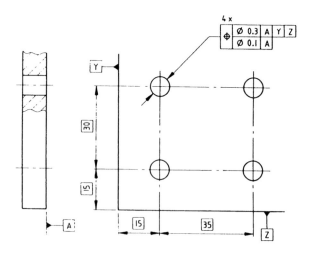

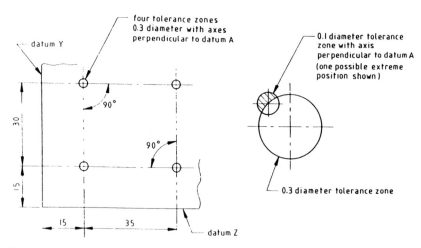

Figure 21.19: Composite positional tolerances according to BS 308; the actual axis may utilize the hatched zone.

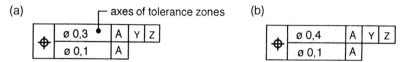

Figure 21.20: Positional tolerances: (a) applied to the centres of (other) positional tolerance zones; (b) applied to the actual axes only.

21.3 DIN 7167 AND FORMER PRACTICES ACCORDING TO DIN 406 AND DIN 7182

The International Standards ISO 1101, ISO 5458, ISO 5459, ISO 3040, ISO 2692 and ISO 8015 have been adopted as DIN Standards without any alteration.

In addition, DIN 7167 has been issued. It specifies the principle of dependency for drawings produced according to DIN Standards without an indication of the principle of independence (without indication of ISO 8015 or DIN 2300). This applies to size and form of isolated features only, as described in 17.6 (not to related features). It does not apply to features to which a form tolerance greater than the size tolerance applies (e.g. to stock material).

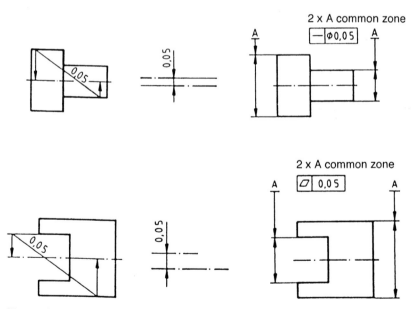

Figure 21.21: Permissible eccentricity according to DIN 406 June 1968 and the equivalent indication according to ISO 1101.

The former German Standard DIN 406 had standardized a drawing indication as shown in Figure 21.21. It specified the maximum permitted separation of axes or median faces. The equivalent indication according to ISO 1101 is also shown in Figure 21.21.

The former German Standard DIN 7182 Part 4 March 1959 symbolized the permissible roundness deviation or the permissible cylindricity deviation by word indication as shown in Figure 21.22. In this case the tolerance was diameter-related (not radius-related as in ISO 1101). The equivalent indication according to ISO 1101 is also shown in Figure 21.22.

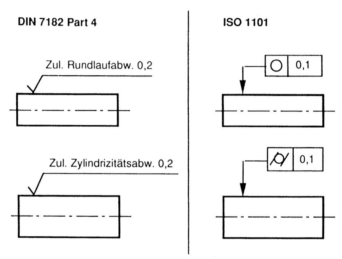

DIN 7182 Part 4 | **ISO 1101**

Zul. Rundlaufabw. 0,2

Zul. Zylindrizitätsabw. 0,2

Figure 21.22: Permissible roundness and cylindricity deviations according to DIN 7182 Part 4 March 1959 and the equivalent indication according to ISO 1101.

21.4 COMECON STANDARDS

According to the Comecon Standards ST RGW 368 -76, distinctions are to be made between radius-related and diameter-related tolerance indications (Table 21.4).

According to ISO 1101 and to the standards of western countries (e.g. ANSI Y14.5M and BS 308 P.3), tolerances of orientation or location also limit the form deviations of the toleranced features (surfaces or axes or median faces). These standards specify that the actual surface or the actual axis or actual median face should be contained in the tolerance zone.

Table 21.4: Tolerance indications according to Comecon and ISO Standards

Tolerance	Former practice RS 430-65 TGL 19085	ST RGW 368-76 TGL 31049 radius-related	ST RGW 368-76 diameter-related	ISO 1101:1983 diameter-related
Coaxiality	⌐ 0,1	◎ R 0,1	◎ ϕ 0,2	◎ ϕ 0,2
Symmetry	÷ 0,1	≡ T/2 0,1	≡ T 0,2	≡ 0,2
Position of axis	+ 0,1	⊕ R 0,1	⊕ ϕ 0,2	⊕ ϕ 0,2
Positional of plane or straight line	—	⊕ T/2 0,1	⊕ T 0,2	⊕ 0,2
Crossing of two axes	✕ 0,1	✕ T/2 0,1	✕ T 0,2	—

This does not apply according to the Comecon Standards ST RGW 301-76 and ST RGW 368 -76. These specify that the orientation or location tolerance zone applies to the geometrical ideal contacting element (Figure 18.14) or the axis of the geometrical ideal contacting cylinder. The contacting element or contacting cylinder is orientated according to the minimum-rock requirement. With holes, the contacting cylinder is the maximum inscribed cylinder while with shafts, it is the minimum circumscribed cylinder.

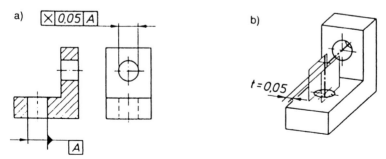

Figure 21.23: Crossing tolerance of two axes: (a) drawing indication; (b) tolerance zone.

(a)   (b)

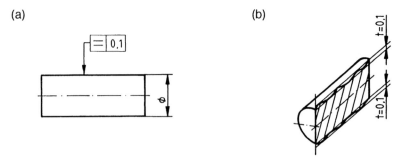

Figure 21.24: Tolerance of the longitudinal section profile: (a) drawing indication; (b) tolerance zone.

According to ST RGW 368 -76, there is a crossing tolerance of two axes with the symbol X. Figure 21.23 shows the drawing indication and the relevant tolerance zone.

According to ST RGW 368 -76, there is a tolerance of the longitudinal section profile with the symbol =. Figure 21.24 shows the drawing indication and the relevant tolerance zone.

This tolerance considers generator lines of cylindrical features in sections containing the axis of the contacting cylinder. In these sections the tolerance zones are composed of two parallel pairs of parallel straight lines each the tolerance value apart. For measurement the pair of parallel straight lines should be orientated in such a way that the greater of the two distances within the pairs of parallel straight lines becomes a minimum.

With the deviation of the longitudinal section profile, the cylindricity deviation can be assessed by a close approximation. According to Heldt (see Ref. [12]), the cylindricity deviation δ_z is approximately equal to the roundness deviation δ_r plus the longitudinal section profile deviation δ_q:

$$\delta_z \approx \delta_r + \delta_q.$$

22 Synopsis of ISO Standards

The standards on Geometrical Product Specifications GPS have been developed by different ISO Technical Committees and only when there was a strong demand for them. Therefore in the field of GPS some standards are still missing or not complete. Table 22.1 gives a synopsis of the ISO Standards that exist or are in preparation.

Columns 1–6 concern the following aspects:

(1) drawing indications (symbolizations);

(2) definition of tolerances (e.g. tolerance zones);

(3) definition of actual feature characteristics or parameters;

(4) assessment of workpiece deviations, and comparison with tolerance limits;

(5) measurement equipment requirements;

(6) calibration requirements.

ISO has formed a Joint Harmonization Group (ISO TC 3 - 10 - 57/JHG) composed of delegates from the relevant ISO Committees. The tasks of this group are to identify gaps and, if existing, contradictions in the standards and to develop future harmonized standards. Typical aims are standards on

- definition of actual size (e.g. of cylindrical features, conical features, features composed of two parallel faces);

- definition of actual distances (e.g. actual axes, stepped faces);

- definition of actual radius;

- definition of actual derived features (actual axes, actual median faces);

- definition of filter requirements in order to distinguish between roughness, waviness, form, orientation and location, size and distance;

- identification of calibration requirements;

- possibly also requirements for measurement strategy and assessment of measurement uncertainties.

Synopsis of ISO standards

Chain link number			1	2
Geometrical charac-teristic of feature	Geometric sub-characteristic of feature or parameters		Product documenta-tion indication - Codifi-cation	Definition of toleran-ces - Theoretical definition and values
Size			ISO 129, ISO 406, ISO-286-1	ISO286-1, ISO286-2, ISO1829
Distance	"Step" distance (height)		ISO129, ISO406	
	Distance between real or derived feature and derived feature		ISO129, ISO406	
Radius			ISO129	
Angle (tolerance in degrees)	Angle between real features		ISO129, ISO1119 (R)	
	Angle between real or derived and derived feature		ISO129	
Form of line indepen-dent of datum	Real feature (line)	Profile any line	ISO1101 (R), ISO1660	ISO1101 (R)
		Straightness	ISO1101 (R)	ISO1101 (R), ◻ISO12780-1
		Roundness	ISO1101 (R)	ISO1101 (R), ISO6318 (W)
	Derived feature (line)	Profile any line	ISO1101 (R), ISO1660	ISO1101 (R)
		Straightness	ISO1101 (R)	ISO1101 (R)
		Roundness	ISO1101 (R)	ISO1101 (R)
Form of line dependant of datum	Real feature (profile of any line)		ISO1101 (R), ISO1660	ISO1101 (R)
	Derived feature (profile of any line)		ISO1101 (R), ISO1660	ISO1101 (R)
Form of surface inde-pendent of datum	Real feature	Profile any surface	ISO1101 (R), ISO1660	ISO1101 (R), ISO1660
		Flatness	ISO1101 (R),	ISO1101 (R)
		Cylindricity	ISO1101 (R)	ISO1101 (R)
		Cones	ISO3040	ISO3040
	Derived feature	Profile any surface	ISO1101 (R)	ISO1101 (R)
		Flatness	ISO1101 (R)	ISO1101 (R)
Form of surface depen-dant of datum	Real feature	any surface	ISO1101 (R), ISO1660	ISO1101 (R), ISO1660
		Cones	ISO3040	ISO3040
	Derived feature		ISO1101 (R)	ISO1101 (R)

◻ = ISO standard in progress (WD, CD or DIS)   XXXYY = ISO number not yet known (numbered YY)   (R) = Revision in progress

3	4		5	6
Definitions for actual feature-characteristic or parameter	Assessment of the deviations of the workpiece - Comparison with tolerance limits		Measurement equipment requirements	Calibration requirements - Calibration standards
ISO/R1938 (R)	Limit gauges	ISO/R1938 (R)	ISO/R1938 (R)	ISO/R1938 (R), IS-O3670 (R), ISO-7863, ISO8512-1, ISO8512-2
	Indicating measuring instruments	ISO/R1938 (R)	ISO463 (R), ISO3599 (R), ISO3611, ◻ISO9121, ISO6906 (W) (R), ◻ISO9493, ◻ISO10360-1, -2, ◻-3, ◻-4, ◻ISO13365, ◻XXX01, ◻XXX09, ◻XXX19	ISO3650 (R), ISO-7863, ◻XXX18
			ISO3599 (R), ISO6906 (R), ◻ISO/TR10360-1, ISO10360-2, ◻-3, ◻-4, ◻ISO13365	
			◻ISO/TR10360-1, ISO10360-2, ◻-3, ◻-4, (◻ISO13365)	
			◻ISO/TR10360-1, ISO10360-2, ◻-3, ◻-4	
			◻ISO/TR10360-1, ISO10360-2, ◻-3, ◻-4	
			◻ISO/TR10360-1, ISO10360-2, ◻-3, ◻-4	
ISO1101 (R), ISO-1660	(ISO/TR5460)		◻ISO/TR10360-1, ISO10360-2, ◻-3, ◻-4	
ISO1101 (R), ◻-ISO12780-1	(ISO/TR5460), ◻ISO12-780-2		ISO463 (R), ◻ISO9493, ISO8512, ◻ISO/TR10360-1, ISO10360-2, ◻-3, ◻-4, ◻ISO12780-3, XXX19	◻ISO12780-4
ISO1101 (R), ◻IS-O12181-1	(ISO/TR5460), ◻ISO12-181-2		ISO4291 (W), ISO4292 (W), ISO463 (R), ◻ISO/TR10360-1, ISO10360-2, ◻-3, ◻-4, ◻ISO12181-3, ◻ISO13365	◻ISO12181-4
	(ISO/TR5460)		◻ISO/TR10360-1, ISO10360-2, ◻-3, ◻-4	
	(ISO/TR5460)		◻ISO/TR10360-1, ISO10360-2, ◻-3, ◻-4	
			◻ISO/TR10360-1, ISO10360-2, ◻-3, ◻-4	
ISO1101 (R), ISO-1660	(ISO/TR5460)		◻ISO/TR10360-1, ISO10360-2, ◻-3, ◻-4	
	(ISO/TR5460)		◻ISO/TR10360-1, ISO10360-2, ◻-3, ◻-4	
ISO1101 (R)	(ISO/TR5460)		◻ISO/TR10360-1, ISO10360-2, ◻-3, ◻-4	
ISO1101 (R), ◻IS-O12781-1	(ISO/TR5460), ◻ISO12-781-2		ISO463 (R), ◻ISO9493, ISO8512, ◻ISO/TR10360-1, ISO10360-2, ◻-3, ◻-4, ◻ISO12781-3, XXX19	◻ISO12781-4
ISO1101 (R9, ◻IS-O12180-1	(ISO/TR5460), ◻ISO12-180-2		◻ISO/TR10360-1, ISO10360-2, ◻-3, ◻-4, ◻ISO12180-3	◻ISO12180-4
			◻ISO/TR10360-1, ISO10360-2, ◻-3, ◻-4, ISO3611, ◻ISO13365	
	(ISO/TR5460)		◻ISO/TR10360-1, ISO10360-2, ◻-3, ◻-4	
	(ISO/TR5460)		◻ISO/TR10360-1, ISO10360-2, ◻-3, ◻-4	
ISO1101 (R)	(ISO/TR5460)		◻ISO/TR10360-1, ISO10360-2, ◻-3, ◻-4	
			◻ISO/TR10360-1, ISO10360-2, ◻-3, ◻-4, ◻ISO13365	
	(ISO/TR5460)		◻ISO/TR10360-1, ISO10360-2, ◻-3, ◻-4	

(W) = To be with drawn   (Parenthesis) = Not sufficient definitions

Synopsis of ISO standards

Chain link number		1	2
Geometrical characteristic of feature	Geometric sub-characteristic of feature or parameters	Product documentation indication - Codification	Definition of tolerances - Theoretical definition and values
Orientation	Real feature (line or plane) — Parallelism (0°)	ISO1101 (R)	ISO1101 (R)
	Real feature (line or plane) — Perpendicularity (90°)	ISO1101 (R)	ISO1101 (R)
	Real feature (line or plane) — Angularity	ISO1101 (R)	ISO1101 (R)
	Derived feature — Parallelism (0°)	ISO1101 (R), ISO10578	ISO1101 (R), ISO10578
	Derived feature — Perpendicularity (90°)	ISO1101 (R), ISO10578	ISO1101 (R), ISO10578
	Derived feature — Angularity	ISO1101 (R), ISO10578	ISO1101 (R), ISO10578
Location	Real feature — Position	ISO1101 (R)	ISO1101 (R), ISO5458 (R), ISO10578
	Derived feature — Position	ISO1101 (R), ISO10578	ISO1101 (R), ISO5458 (R), ISO10578
	Derived feature — Coaxiality	ISO1101 (R), ISO10578	ISO1101 (R), ISO10578
	Derived feature — Concentricity	ISO1101 (R), ISO10578	ISO1101 (R), ISO10578
	Derived feature — Symmetry	ISO1101 (R), ISO10578	ISO1101 (R), ISO10578
Circular run out		ISO1101 (R)	ISO1101 (R)
Total run out		ISO1101 (R)	ISO1101 (R)
Datums	Datums — Datums associated with real features	ISO1101 (R), ISO5459	(ISO5459)
	Datums — Datums associated with derived features	ISO1101 (R), ISO5459	(ISO5459)
	Datum targets	ISO1101 (R), ISO5459	(ISO5459)
	Datum systems	ISO1101 (R), ISO5459	(ISO5459)
Surface roughness	M-System - Ra, Ry	ISO1302	ISO4287/1 (R), ISO4287/2, ISO468 (W)
	Motif method - R		ᴏISO12085
	S, Sm, tp		ISO4287-1 (R), ISO468 (W)
	Rk, Rpk, Rvk, Rm1k, Rm2k		ᴏISO13365-1, ᴏISO13365-2
	Rpq, Rqv, Rmq		ᴏISO13365-1, ᴏISO13365-3
	Surface 3D characteristic		
Surface waviness	M-system - Wa, Wy		ᴏISO10479 (W), ISO4287-1 (R)
	Motif method - W		ᴏISO10479 (W), ISO4287-1 (R)
Surface defects		ᴏISO8785	ᴏISO8785
Edges		ᴏISO10135-2	ᴏISO10135-2

ᴏ = ISO standard in progress (WD. CD or DIS)   XXXYY = ISO number not yet known (numbered YY)   (R) = Revision in progress

398

3	4	5	6
Definitions for actual feature-characteristic or parameter	Assessment of the deviations of the workpiece - Comparison with tolerance limits	Measurement equipment requirements	Calibration requirements - Calibration standards
ISO1101 (R)	(ISO/TR5460)	ISO8512-1, -2, □ISO/TR10360-1, ISO10360-2, □-3, □-4	
ISO1101 (R)	(ISO/TR5460)	□ISO/TR10360-1, ISO10360-2, □-3, □-4	
ISO1101 (R)	(ISO/TR5460)	□ISO/TR10360-1, ISO10360-2, □-3, □-4	
	(ISO/TR5460)	□ISO/TR10360-1, ISO10360-2, □-3, □-4	
	(ISO/TR5460)	□ISO/TR10360-1, ISO10360-2, □-3, □-4	
	(ISO/TR5460)	□ISO/TR10360-1, ISO10360-2, □-3, □-4	
ISO1101 (R)	(ISO/TR5460)	□ISO/TR10360-1, ISO10360-2, □-3, □-4	
	(ISO/TR5460)	□ISO/TR10360-1, ISO10360-2, □-3, □-4	
	(ISO/TR5460)	□ISO/TR10360-1, ISO10360-2, □-3, □-4	
	(ISO/TR5460)	□ISO/TR10360-1, ISO10360-2, □-3, □-4	
	(ISO/TR5460)	□ISO/TR10360-1, ISO10360-2, □-3, □-4	
ISO1101 (R)	(ISO/TR5460)	ISO463 8R), □ISO9493, □ISO/TR10360-1, ISO10360-2, □-3, □-4, XXX19	
ISO1101 (R)	(ISO/TR5460)	ISO463 8R), □ISO9493, □ISO/TR10360-1, ISO10360-2, □-3, □-4, XXX19	
(ISO5459), XXX26	(ISO/TR5460), XXX27	ISO8512-1, -2, □ISO/TR10360-1, ISO10360-2, □-3, □-4, XXX28	
(ISO5459), XXX26	(ISO/TR5460), XXX27	□ISO/TR10360-1, ISO10360-2, □-3, □-4, XXX28	
(ISO5459), XXX26	(ISO/TR5460), XXX27	ISO8512-1, -2, □ISO/TR10360-1, ISO10360-2, □-3, □-4, XXX28	
(ISO5459), XXX26	(ISO/TR5459), XXX27	□ISO/TR10360-1, ISO10360-2, □-3, □-4, XXX28	
ISO4288 (R)	ISO4288 (R), 2632-1 (W), -2 (W)	ISO1878, ISO1879 (W), ISO1880 (W), ISO3274 (R), □ISO11562, ISO2632-1 (W), -2 (W)	ISO2632-1 (W), -2 (W), ISO5436 (R), □ISO12179
□ISO12085			
		ISO1880 (W), ISO3274 (R), □ISO11562	ISO5436 (R), □ISO12179
		ISO1880 (W), ISO3274 (R), □ISO11562	ISO5436 (R), □ISO12179
□ISO13423, ISO4287-1 (R)		ISO1880 (W), ISO3274 (R), □ISO11562	ISO5436 R, □ISO12179
□ISO12085, ISO4287-1 (R)			

(W) = To be with drawn   (Parenthesis) = Not sufficient definitions

Table 22.1, entitled The General GPS (Geometrical Product Specification) Chains of Standards Matrix, has been developed by ISO/TC 3 - 10 - 57/JHG under the leadership of the Danish Standards Association (DSA), Baunegaardsvej 73, DK-2900 Hellerup, Denmark, and will be amended continuously. The latest table may be purchased from the national standardization body (e.g. ANSI in the USA, AFNOR in France, BSI in the UK, DIN in Germany and DSA in Denmark) or from the author.

This standardization work has now become urgent, with current efforts to eliminate the trade barriers and with the growth of worldwide industrial cooperation. It is necessary, for example, to give unambiguous definitions for the programming of coordinate measuring machines and unambiguous rules for the traceability of measurements according to ISO 9001.

Standards

ISO 286-1	ISO System of limits and fits: bases of tolerances, deviations and fits
ISO 1101	Geometrical tolerancing
ISO 1660	Dimensioning and tolerancing of profiles
ISO 2768-2	General geometrical tolerances
ISO 2692	Maximum material requirement, least material requirement, reciprocity requirement
ISO 3040	Dimensioning and tolerancing of cones
ISO 4291	Methods for the assessment of departures from roundness: measurement of variations in radius
ISO 4292	Methods for the assessment of departures from roundness: measurement by two- and three-point methods
ISO 5458	Geometrical tolerancing: positional tolerancing
ISO 5459	Datums and datum systems for geometrical tolerancing
ISO 6318	Measurement of roundness: terms, definitions and parameters of roundness
ISO 7083	Symbols for geometrical tolerancing: proportions and dimensions
ISO 8015	Fundamental tolerancing principle
ISO 10 579	Dimensioning and tolerancing: non-rigid parts
ISO 10 360-1	Coordinate metrology, Part 1: definitions and applications of the fundamental geometrical principles
ISO TR 5460	Geometrical tolerancing: — tolerancing of form, orientation, location and run-out; — Verification principles and methods, guidelines
ANSI Y14.5M-1982	Dimensioning and tolerancing
ANSI B89.3.1-1972	Measurement of out-of-roundness
BS 308 P.3	Engineering drawing practice, geometrical tolerancing
BS 3730	Assessment of departures from roundness
BS 7172	British Standard Guide to assessment of position, size and departure from nominal form of geometrical features
DIN 4760	Form deviation, waviness, surface roughness; system of order, terms and definitions
DIN 6784	Edges of workpieces: terms, drawing indications
DIN 7167	Relationship between dimensional tolerances and form and parallelism tolerances; envelope requirement without indication

Standards

DIN 7184	Geometrical tolerances; definitions and drawing indications (superseded by DIN ISO 1101)
DIN 7186 T.1	Statistical tolerancing: definitions, applications, drawing indications
DIN 8570 T.3	General geometrical tolerances for welded parts
DIN 40680 T.2	General form tolerances for ceramic parts in electrical application
VDI/VDE 2601 T.1	Requirements on the surface structure to cover function capability of surfaces manufactured by cutting; list of parameters
TGL 39 092	Methods of measuring geometrical deviations: general principles
TGL 39 093	Methods of measuring deviations from straightness
TGL 39 094	Methods of measuring deviations from flatness
TGL 39 095	Methods of measuring deviations from parallelism
TGL 39 096	Methods of measuring deviations from roundness
TGL 39 097	Methods of measuring deviations from cylindricity
TGL 39 098	Methods of measuring deviations of the longitudinal section profile
TGL 43 041	Methods of measuring straightness deviations of axes
TGL 43 042	Methods of measuring deviations from coaxiality
TGL 43 043	Methods of measuring the radial run-out deviations
TGL 43 044	Methods of measuring the axial run-out deviations
TGL 43 045	Methods of measuring the run-out deviations in a given direction
TGL 43 529	Methods of measuring the radial total run-out deviations
TGL 43 530	Methods of measuring the axial total run-out deviations
ST RGW 301-76	Geometrical tolerances: fundamental terms
ST RGW 368-76	Geometrical tolerances: drawing indications

Bibliography

[1] Bänninger, E. Das Maximum-Material-Prinzip. *VSM/SNV Norm. Bull.* **10** (1965) 109–120.

[2] Kirschling, G. *Qualitätssicherung und Toleranzen* (Berlin: Springer-Verlag, 1988).

[3] Smirnow, N.W. *Mathematische Statistik in der Technik* (Berlin: VEB Deutscher Verlag der Wissenschaften, 1969).

[4] Trumpold, H., Beck, C. and Riedel, T. *Tolerierung von Maßen und Maßketten im Austauschbau* (Berlin: VEB-Verlag Technik, 1984).

[5] Wirtz, A. Die theoretischen Grundlagen der Dreikoordinatenmeßtechnik. *Technische Rundschau* **38** (1985) 12–23.

[7] Warnecke, H.J. and Dutschke, W. *Fertigungsmeßtechnik* (Berlin: Springer-Verlag, 1984).

[8] Lotze, W. Form- und Lageabweichungen definitionsgemäß mit Koordinatenmeßgeräten prüfen (mit umfangreichem Literaturverzeichnis). *Technische Rundschau* **31** (1990) 36–45.

[9] Neumann, H.J. *Koordinatenmeßtechnik* (Landsberg/Lech: Verlag Moderne Industrie, 1990).

[10] Barth, U. Kombinierte Zweipunkt-Dreipunkt-Messung zum gleichzeitigen Bestimmen von Maß- und Kreisformabweichungen. *Feingerätetechnik* **33** (1984) 21–23.

[11] Wirtz, A. and Maduda, M. Fertigungsgerechte Rundheitsmessung. *Fertigung* **5** (1972) 127–137.

[12] Heldt, E. Komplexe Bestimmungsgröße Zylinderform und der Zusammenhang mit ihren Komponenten. *Feingerätetechnik* **33** (1984) 10–14.

[13] Trumpold, H. and Richter, G. Unterschiede zwischen DDR-Standards und DIN-Normen auf dem Gebiet der Form- und Lageabweichungen. *Feingerätetechnik* **39** (1990) 451–453.

Index

Index